Table of Contents

List of Figures

Figures

Tables

Preface

This study was conducted for Dr. Jeff Holland, Director of the Engineer Research and Development Center under a project called, "Integrated Risk Management for Climate Change," via the Center Directed Research Program. The technical reviewer was Dr. Todd Bridges, CERD-EM-D.

The work was performed by the Environmental Processes Branch (CN-N) of the Facilities Division (CF), Construction Engineering Research Laboratory (CERL). The CERL Principal Investigator was Dr. James Westervelt. The modeling groups, the Program for Climate Model Diagnosis and Intercomparison (PCMDI) and the WCRP's Working Group on Coupled Modelling (WGCM) are acknowledged for their roles in making available the WCRP CMIP3 multi-model dataset. Support of this dataset is provided by the Office of Science, US Department of Energy. William Meyer is Chief, CEERD-CN-N, and Dr. John Bandy is Chief, CEERD-CF. The Director of ERDC-CERL is Dr. Ilker R. Adiguzel.

CERL is an element of the US Army Engineer Research and Development Center (ERDC), US Army Corps of Engineers. The Commander and Executive Director of ERDC is COL Kevin J. Wilson, and the Director of ERDC is Dr. Jeffery P. Holland.

Conversion Factors

Non-Systeme Internationale (SI) units of measurement used in this report can be converted to SI units as follows:

Multiply	By	To Obtain
acres	4,046.873	square meters
cubic feet	0.02831685	cubic meters
cubic inches	0.00001638706	cubic meters
degrees (angle)	0.01745329	radians
degrees Fahrenheit	(5/9) x (°F – 32)	degrees Celsius
degrees Fahrenheit	(5/9) x (°F – 32) + 273.15	kelvins
feet	0.3048	meters
gallons (US liquid)	0.003785412	cubic meters
horsepower (550 ft-lb force per second)	745.6999	watts
inches	0.0254	meters
kips per square foot	47.88026	kilopascals
kips per square inch	6.894757	megapascals
miles (US statute)	1.609347	kilometers
pounds (force)	4.448222	newtons
pounds (force) per square inch	0.006894757	megapascals
pounds (mass)	0.4535924	kilograms
square feet	0.09290304	square meters
square miles	2,589,998	square meters
tons (force)	8,896.443	newtons
tons (2000 pounds, mass)	907.1847	kilograms
yards	0.9144	meters

1 Introduction

1.1 Background

According to the 2007 *Fourth Assessment Report* by the Intergovernmental Panel on Climate Change (IPCC 2007), global surface temperature increased 0.6 ± 0.2 °C (1.08 ± 0.36 °F) during the 20th century (IPCC 2007a). Much of the observed temperature increase since the middle of the 20th century has been caused by increasing concentrations of greenhouse gases, which result from human activity such as the burning of fossil fuel and deforestation. Global dimming, a result of increasing concentrations of atmospheric aerosols that block sunlight from reaching the surface, has partially countered the effects of warming induced by greenhouse gases.

Climate model projections summarized in the 2007 IPCC report indicate that global surface temperature is likely to rise between 1.1 and 6.4 °C (2.2 and 11.5 °F) during the 21st century (IPCC 2007a).

In February 2010, the President's Council on Environmental Quality (CEQ) issued draft guidance to all Federal agencies concerning the manner in which climate change should be included in the evaluation of environmental effects under the National Environmental Policy Act (NEPA) (Secretary of Defense 2010). Under NEPA, Federal agencies are required to evaluate the environmental impacts of proposed Federal actions and, wherever possible, to explore a broad range of options for minimizing potentially adverse outcomes and consequences that are caused – wholly or in part – by those actions. This new guidance extends the issues to be considered to include greenhouse gas (GHG) emissions and climate change, and how agencies should address the interactions between their proposed actions and these factors. Specifically, the guidance states that:

> With regard to the effects of climate change on the design of a proposed action and alternatives, Federal agencies must ensure the scientific and professional integrity of their assessment of the ways in which climate change is affecting or could affect environmental effects of the proposed action ...
> Climate change can increase the vulnerability of a resource, ecosystem, or human community, causing a proposed action to result in consequences that are more

damaging than prior experience with environmental impacts analysis might indicate. ...

Agencies should consider the specific effects of the proposed action (including the proposed action's effect on the vulnerability of affected ecosystems), the nexus of those effects with projected climate change effects on the same aspects of our environment, and the implications for the environment to adapt to the projected effects of climate change....

Where agencies consider climate change modeling to be applicable to their NEPA analysis, agencies should consider the uncertainties associated with long-term projections from global and regional climate change models.

As with other agencies, the effects of climate change are expected to impact Continental United States (CONUS) military installations. In particular, Army installations have large land based range areas used for testing, training or maneuvers. Climate change has the potential to affect at least three concerns of most interior continental installations:

- erosion characteristics
- the management of Threatened and Endangered Species (TES).
- the appearance and increase of noxious invasive species.

This study is an initial evaluation and analysis of data already at hand. There is no question that the application of more time and resources could result in a better detailed evaluation of climate affects. We were disappointed that a USEPA report dealing with ecological changes on public lands (US Climate Change Science Program 2008) did not include the Department of Defense (DoD) as one of the participating agencies. However, a study funded by the DoD Strategic Environmental Research and Development Program (SERDP) examines many of the same problems we highlight in this report (Smith et al. 2010).

The US Army Corps of Engineers Engineering Research and Development Center (ERDC) has set aside a portion of funding from The Center Directed Research Program for the ERDC *Framework for Assessing the Environmental Effects of Climate Change for the Military* to build capability and research capacity focused on military installation management.*

---

* Much of the following discussion is taken directly from the research proposal, the funds from which support this portion of the research initiative: Proposal CDR SOW 3-1-10 Integrated Modeling and Risk Analysis for the Environmental Consequences of Climate Change: *A Framework for Assessing the Envi-*

ERDC is pursuing five major tasks as a part of the parent project from which this report was derived:

1. Design of Analytical System Architecture
2. Climate Downscaling, Calibration, and Integration with Consequence Models
3. Hydrologic Impacts of Climate Change
4. Development of Ecological Process Models
5. Development of Integrated Risk and Decision Analysis Framework.

The research in this report represents an initial action for Task 4: Development of Ecological Process Models. Specifically, we identify the Army installations that show the greatest risk of severe effects due primarily to climate change. Most literature examining military installations has dealt with the effect of rising sea levels on coastal areas, a concern more important to Navy and Marine interests than for Army installations. Land managers deal with their lands in the context of the ecosystem in which they reside. What happens if that ecosystem changes? How does a land manager then care for those changing lands while still supporting his/her mission?

1.2 Objective

The objective of this report is to make an initial, broad evaluation of the effects of forecasted climate change on ecosystems and related concerns (including erosion, TES, and invasive species). We will then apply our evaluation to CONUS Army installations, rank order the impacted installations and provide insight as to the probable changes operation that will be required for the installations to carry on their responsibilities and missions.

1.3 Approach

We provide a background understanding of the Department of Defense (DoD) and Army documents and procedures relating to climate change issues in Chapter 2 and a review of historic ecosystem characterizations in Chapter 3. Chapter 4 consists of a broad review of climate change research — particularly the predicted spatial distribution of expected changes. In Chapter 5, we describe the data sets and the procedures used in our evalu-

ronmental Effects of Climate Change for the Military. (Statement of Work for, US Army Corps of Engineers, Engineering Research and Development Center, 1 March, 2010).

ation of climate change, ecosystems, erosion, TES and invasive species affects. Time horizons for these data sets are the year 2000 and 2099. Finally, in Chapter 6 we provide a rank ordering of the climate affects on over 100 Army installations, compare the rankings and draw preliminary conclusions.

This report addresses with the issues of:

- precipitation change under three different climate scenarios, broken down by installation
- temperature change under three different climate scenarios broken down by installation
- ecological changes for
  - Bailey's ecosystem characterization by installation
  - Omernik's ecosystem characterization by installation
  - US Geological Survey (USGS) Gap Analysis Program (GAP) ecosystem characterization by installation
- effects on erosion broken down by installation
- general effects on TES
- general effects on invasive species.

1.4 Scope

This investigation reviews the available literature that specifically supports the spatial distribution of climate change predictions. We do not attempt to generate new predictions on our own. We also assume that the military missions at installations will remain the same as they are today, which is unlikely to be universally true over the time horizon we used (2000-2099). This is a preliminary report concerning general affects of climate change on ecosystems across CONUS. We look at individual installations to generate a rank order listing of affects where possible. It is clear the next step needs to be a more detailed evaluation of effects within regions and at specifically identified installations.

Some may claim that ecosystems are a poor metric for monitoring climate change, but military lands are managed within an ecosystem context rather than precipitation and temperature. Consequently, this initial study is intended to make the connection between climate change research and ecosystems to suggest what circumstances DoD managers will have to deal with in the future. Also, in general, it is difficult for individuals to visualize the effects of temperature and precipitation changes, but it is easy to understand that the plants and animals you will see outside your window are about to change. Ecosystem response will lag decades behind actual climate shifts, but will eventually reconfigure in the manner suggested here. Global Climate Model (GCM) outputs of temperature and precipitation are the better metrics, but other professionals are researching that area; furthermore, the concern of our research are military installations, which makes a more holistic ecosystem evaluation much more valuable. We want it be clear that we are not using ecosystems as a metric for climate change itself, but simply exploring the potential effects of an altered climate. Therefore, in this study we take data from those who predict climate characteristics and apply those metrics to the likely response of ecosystems and other natural processes.

1.5   Mode of technology transfer

This report will be made accessible through the World Wide Web (WWW) at URL: http://www.cecer.army.mil

2 Climate Change and the Military

2.1 The Quadrennial Defense Review

In February 2010, the Quadrennial Defense Review (QDR, see Figure 1) was the first Department of Defense (DoD) publication to address the issue of the Growing Need to Consider Risks and Response Strategies for Climate Change. In the QDR, the DoD explicitly acknowledges that climate change will likely affect the nature and scope of future missions, as well as training and testing assets of military installations. Specifically it says the military must:

- reliably assess the causes and consequences of climate change
- arrive at a coherent and robust understanding of a broad range of possible response options that *minimize adverse environmental consequence* and *maximize the likelihood of mission success* around the globe.

The QDR deals with general military concerns for climate change. Nothing is directly said about Army lands management impacts or Army installation mission impacts due to climate change. Many of the references we could find in our literature review deal with climate change on military installations largely in terms of sea level rise resulting from ice melt. This is an impact on Navy yards and Marine installations certainly, but occurs at Army installations rarely. Most Army installations are large occasionally conglomerated associations of lands in an inland location and very often in the drier Western areas of the Western United States.

2.2 Framework for assessing the environmental effects of climate change for the military

Considerations dealing with climate change prediction presuppose an analysis and evaluation capability that is far from trivial in design, scope, and purpose. In the military realm, for example, *mission*, *geo-physical space/terrain*, and *human agency* are tightly interwoven. As a consequence, efforts to inform military decisions about the prospect of climate change require a unique set of analytic capabilities, incorporating an understanding of climate change (at various levels of aggregation), weather, watershed processes, ecological impacts, and landscape evolution. Risk and uncertainty are endemic features of the climate change problem; consequently, decisionmakers require sophisticated tools for effectively managing risk as part of their decision evaluation and implementation processes.

The QDR and the climate change guidance released by CEQ speak to the need for developing a scientifically rigorous approach for producing credible forward-looking projections examining potential influence of climate change on ecosystems. Further, there is a need for a risk analytic framework to guide the required assessments, modeling, and evaluation of alternative response strategies.

The overarching goal of the ecological modeling component of this project is to forecast the consequences of climate change on the ecosystems of military installations and the surrounding region. The specific modeling that is accomplished will follow directly from the team's identification of the area of interest, the specific questions of regional relevance concerning ecosystem consequences, and the available expertise and data.

2.3   National-scale ecological impact analysis

The impact of climate change will vary from location to location. This report conducts a national survey of installations to identify likely areas that will experience dramatic climate change impacts of relevance to military installations. Ecosystems in CONUS have been categorized, defined, and located through analyses conducted by both Omernik and Bailey. We have compiled national scale maps (developed primarily by the USGS) that will include elevation, soils, historic climate/weather, insulation, elevation, and latitude. We will develop statistical models that correlate specific ecosystem locations with these independent values and apply the models to the national suite of maps. Then, using available climate change forecasts for climate/weather, we will run the model against these scenarios to generate maps suggesting how national-scale ecosystem patterns might shift. The resulting suite of forecast maps will provide the project with a range of forecasts for significant ecosystem change for all locations across the country, including military installations.

In recent years, researchers have begun to explore the manner and degree to which climate change will impact military assets in a diverse range of theatres of operation. For instance, the status and condition of infrastructure, training and testing assets, and natural resources contained within military installations are strongly influenced and dependent on ecosystem and landscape structure and functions. Though the exact timeframe is still somewhat uncertain, climate change is likely to drive changes in ecosystems with consequent effects on infrastructure performance and capacity, training and testing capabilities, natural resource management, and threatened and endangered species.

Army installations develop installation natural resource management plans (INRMPs) and refine them every 5 years. These plans chart the management of the installation from an environmental standpoint out 50 to 100 years into the future. The plans look so far into the future because it takes a long time to develop the types of complex ecological landscapes on which many species rely. To date, INRMPs have not reflected anticipated risks of climate change. It is likely that some installation INRMPs will be dramatically changed in response to local climate change forecasts and their impact on the health of current and planned systems.

2.4 Ecosystem change

Climate change will push existing ecosystems towards thresholds where the current systems will be restructured to the point of replacement with significantly altered or "novel" ecosystems (IPCC 2007a). We are on the threshold of dramatic changes to installation ecosystems (changing species densities, local extirpations, and moving to completely different ecosystems). Currently, military lands are managed to maintain what currently exists (management principle of "stationarity"). Under this paradigm, future land management intensity and costs will increase unless we develop a better understanding of ecosystem transformations in response to climate change. The best opportunity for anticipating the future is to model these processes.

ERDC's extensive ecosystem research on military lands over the past three decades includes examinations of responses of endangered species to military missions, habitat rehabilitation and remediation, invasive species, and effects of land condition on the military's training and testing mission. Climate change can be seen as a global "catalyst" for environmental and ecological stressors that drive change in natural systems. In every region of the globe, ecosystems are experiencing transformations in response to multiple stressors, including climate change, land use change, fragmentation of natural areas, reduced water infiltration and availability, biodiversity loss, and the spread of invasive species. As a result, ecological systems that do not currently exist (systems that are naturally emerging in response to human actions) will continue to develop with unknown consequences for military land-use requirements, ecosystem service provision, and environmental security.

It is acknowledged that not all the ecosystem identifications are those of natural systems. For example, the category "Pine Plantation" used in the USGS GAP analysis is clearly a human encouraged system. Yet everyone recognizes that the category "Pine Plantation" is species specific; that the species that thrive in that system would not if the conditions for their survival were not just right. In economics this concept is called "highest and best use." Highest and best use is the reason the ecosystem is called "Pine Plantation" rather than "Turkey oak-longleaf pine." In this sense, human encouraged categories may better reflect the environmental conditions than naturally generated categories because human simplification limits their distribution, unlike a natural system in which diversity is an important attribute. For example, if the climate in an area such as the Southeastern United States dries up significantly, humans may respond to the new situation by:

1. Discontinuing production of some crops when traditional commercial plants no longer grow well and profitably
2. Substituting crop species that are more appropriate/adopted to the new conditions
3. Relocating their Pine Plantations to a new, more appropriate climate.

In these cases, human forced, single species dependent ecosystems are likely to response to climate change much faster than natural ecosystems. Only more urban related categories will be less affected. Therefore, as climate changes occur, human encouraged "ecosystems" are at least as likely to shift their prime distribution as natural systems.

Current analytical and simulation capabilities are limited in the ability to evaluate climate change effects and response in natural systems of interest to military planners because of: (1) the highly dynamic spatial and temporal characteristics of natural system response to climate change; (2) the spatial and temporal scale of those responses; and (3) the complex biophysical interactions in these natural systems. The ability to forecast and assess alternative scenarios is relatively lagging in the overall climate change research area, and those areas where such research is being attempted are not adequately robust for DoD requirements.

Key questions facing installations in the years to come include:

- Which installations are at greater risk for habitat disruption that is partially or fully driven by climate change?
- Where will disruptions involve ecoregion shifts?
- How are habitats at installations likely to change and when might these changes occur?
- Should installations invest in maintaining current ecosystem states?
- How will sensitive habitats supporting threatened/endangered species change?

Current management and regulatory frameworks for military land management and planning assume "stationarity" of natural systems under consideration, where ecosystem states and processes operate within an unchanging range of variability. Climate change requires a major paradigm shift away from assumptions of stationarity.

A high risk/high pay-off opportunity for ERDC will be to effectively capture the dynamics and complexities of natural system response to climate change that will be relevant for risk analyses and natural resource management planning and implementation for military lands.

Ecosystems are made up of many characteristics, not just temperature and precipitation. Other concerns that determine ecosystems may shift in response to climate change (e.g., slope or soil acidity), but over the next 100 years, most other concerns are essentially stable. Thus, the time horizon used redefines these other concerns as constants, rather than variables. So there is no point in predicting a trend using constants. Taking the reverse tactic, the GCMs only generate changes in temperature and precipitation so there is no predictive source for example for soil acidity. In addition, of those constraints that define an ecosystem, temperature and precipitation are always of highest priority while slope may have an influence it is not primary in all ecosystems (and in fact is in part derived from rainfall in particular). So although our analysis is limited, it is limited to the most important influences.

2.5   Climate downscaling

One of the other disciplinary areas within the ERDC Framework for Assessing the Environmental Effects of Climate Change for the Military Statement of Work deals with "Climate Downscaling." In the research for this report, we found that the work done in Climate Downscaling provided a good basis for application of GCM results. One of the authors (Dr. John Weatherly, ERDC Cold Regions Research and Engineering Laboratory [CRREL], New Hampshire) provided significant direction and description in allowing an appropriate understanding of the downscaled climatic data used.

3   Historical Ecosystem Characterizations

3.1   Historical ecological characterizations

3.1.1   Bailey's ecoregions

First published in 1983, Dr. Robert G. Bailey's ecoregion classification (Figure 2) became a standard reference in the field of ecology. Dr. Bailey continues to refine this map at the US Department of Agriculture (USDA) Forest Service. Bailey's ecoregions distinguish areas that share common climate and vegetation characteristics. A four-level hierarchy is used to differentiate the ecoregions. *Domains* (the broadest classification) are groups of related climates and are differentiated based on precipitation and temperature. Thus, the domain map is well suited for use in analyzing projected climate change impacts. Four domains are used worldwide and all four appear in the United States: the polar domain, the humid temperate domain, the dry domain, and the humid tropical domain. *Divisions* represent the climates within domains and are differentiated based on precipitation levels/patterns and temperature. Divisions are subdivided into *provinces*, which are differentiated based on vegetation or other natural land covers. Mountainous areas that exhibit different ecological zones based on elevation are identified at the province level. *Sections* are the finest subdivision and are based on terrain features. Much of Bailey's ecoregions were drawn from US Department of the Interior (2011).

Figure 2. Bailey's ecoregions of North America shown at the "Province" level.

3.1.2 Omernik's ecoregions

James M. Omernik developed his ecoregion classifications (Figure 3) while working with the US Environmental Protection Agency's National Health and Environmental Effects Research Laboratory is Corvallis, OR.

The Omernik ecoregion system is based on a four level hierarchy and considers the spatial patterns of both the living and non-living components of the region, such as geology, physiography, vegetation, climate, soils, land use, wildlife, water quality, and hydrology. The patterns affect or reflect differences in ecosystem quality and integrity. All the components are considered when determining the location of ecoregion boundaries, but the relative importance of each component may vary from one ecoregion to another, regardless of the level of the hierarchy. There are four levels in the Omernik ecosystem hierarchy: Level I divides North America into 15 broad ecoregions appropriate for analysis at a global or intercontinental scale.* The 52 Level II ecoregions are useful for national and sub continental overviews of physiography, wildlife, and land use. Level III represents a further subdivision, with 194 ecoregions to describe North America, of which 104 apply to the CONUS. This level is appropriate for regional analysis and decisionmaking; therefore, 84 of the 104 CONUS Level III regions were used in this analysis. Level IV ecoregion identification is underway or complete for most of the United States.

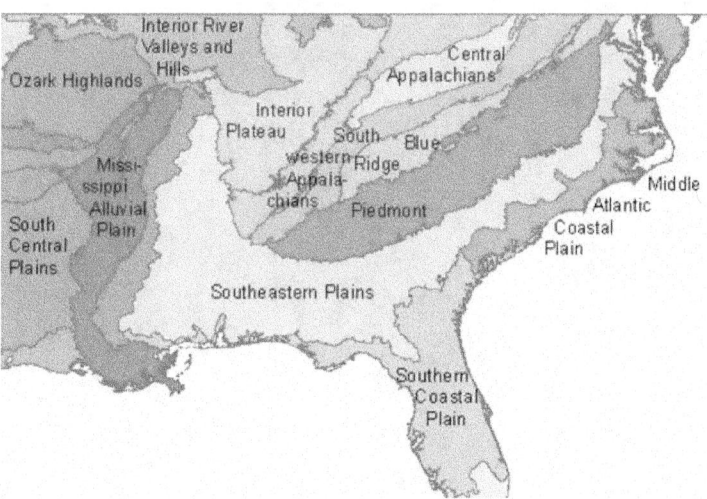

Figure 3. US Ecosystems (Version 3) by Omernik at Level III.

---

* Level I ecoregions were mapped and described by the North American Commission for Environmental Cooperation (CEC) in 1997.

3.2   Recent technologically derived ecological characterizations

3.2.1 USGS Gap Analysis Program (GAP)

One of the more recent notable land cover characterizations was produced under the studies of gaps in species habitats (thus GAP) initiative by John Mosesso, Anne Davidson, and Ron Sepic at the US Geologic Survey. GAP's mission is to keep common species common by providing information on the status of native species. The latest version of the landcover map contains 551 Ecological Systems and modified Ecological Systems; the modified ecological systems represent 32 land use classes that depict developed and/or disturbed land cover classes.

The map in Figure 4 shows data combined from previous GAP projects in the Southwest, Southeast, and Northwest United States with recently updated California data. For areas of the CONUS where ecological system-level GAP data has not yet been developed, data from the LANDFIRE project (aka Landscape Fire and Resource Management Planning Tools) compiled by Landscope (www.landfire.gov) was used. This allows for the construction of a seamless representation of ecological system distributions across the continental United States. This new version is used in our analysis in the same manner as the Bailey's and Omernik's ecosystem maps.

Figure 4. GAP national land cover map.

3.2.2 Hargrove/Hoffman potential multivariate quantitative methods for delineation and visualization of ecoregions

The Hargrove/Hoffman map (Figure 5) of ecosystems is not a well known standard. We were impressed with their paper (Hargrove and Hoffman 2004) because it seemed to hold a great deal of promise for dealing with climate change modeling; consequently, we decided to use their work as part of this research effort. Multivariate clustering is based on fine spatial resolution maps of elevation, temperature, precipitation, soil characteristics, and solar inputs. The coarse ecoregion divisions previously outlined accurately capture intuitively-understood regional environmental differences, whereas the finer divisions in this sort of ecosystem classification highlight local condition gradients, ecotones, and clines.

Figure 5. A Hargrove/Hoffman map. The color of each ecoregion indicates the relative mix of nine environmental conditions inside each ecoregion. Red is "physiographic position" (i.e., low precipitation, high insolation, high elevation, and deep water table). Green is "plant nutrients" (i.e., high soil N, organic matter, and available water). Blue is "temperature" (i.e., few degree-days heat and many degree-days cool). Shown in these Similarity Colors, the borders between individual ecoregions disappear.

Such statistically generated ecoregions can be produced based on user-selected continuous variables, allowing customized regions to be delineated for any specific problem. By creating an objective ecoregion classification, the ecoregion concept is removed from the limitations of human subjectivity, making possible a new array of ecologically useful derivative products. Multiple geographic areas can be classified into a single common set of quantitative ecoregions to provide a basis for comparison, or maps of a single area through time can be classified to portray climate or environmental changes geographically in terms of current conditions.

In the Hargrove data on which we based this research, we were able to access a new data set in preparation for new applications. The new data delineates 30,000 "ecoregions" across the globe. The ecoregions are based on the potentially changing base characterization of the 17 variables listed in Table 1.

Table 1. Hargrove's 17 factors used to define ecosystems in the Multivariate Geographic Clustering (MGC) procedure.

1. Precipitation during the hottest quarter*
2. Precipitation during the coldest quarter*
3. Precipitation during the driest quarter
4. Precipitation during the wettest quarter
5. Ratio of precipitation to potential evapotranspiration
6. Temperature during the coldest quarter
7. Temperature during the hottest quarter
8. Sum of monthly Tavg where Tavg >=5 deg C
9. Integer number of consecutive months where Tavg >= 5 °C (Length of potential growing season)*
10. Available water holding capacity of soil
11. Bulk density of soil
12. Carbon content of soil
13. Nitrogen content of soil
14. Compound topographic index (relative wetness)
15. Solar interception
16. Day/night diurnal temperature difference
17. Elevation (not used in "noelev" clustering)

3.2.3 Other notable ecoregion data sets not used in this study

Technology has advanced and provided researchers with other options for evaluating ecosystem climate change indices. Other means of evaluating change indices nominated to be used in future research studies are illustrated below.

3.2.3.1 Net primary productivity

Net Primary Productivity (NASA 2011) explains how much carbon dioxide vegetation takes in during photosynthesis minus how much carbon dioxide the plants release during respiration (metabolizing sugars and starches for energy). The data come from the Moderate Resolution Imaging Spectroradiometer (MODIS) on the National Aeronautics and Space Administration (NASA) Terra satellite Earth Observing System (EOS) (Figure 6). Values range from near 0 grams of carbon per square meter per day (tan) to 6.5 grams per square meter per day (dark green). A negative value indicates decomposition or respiration overpowered carbon absorption; therefore, more carbon was released to the atmosphere than the plants took in.

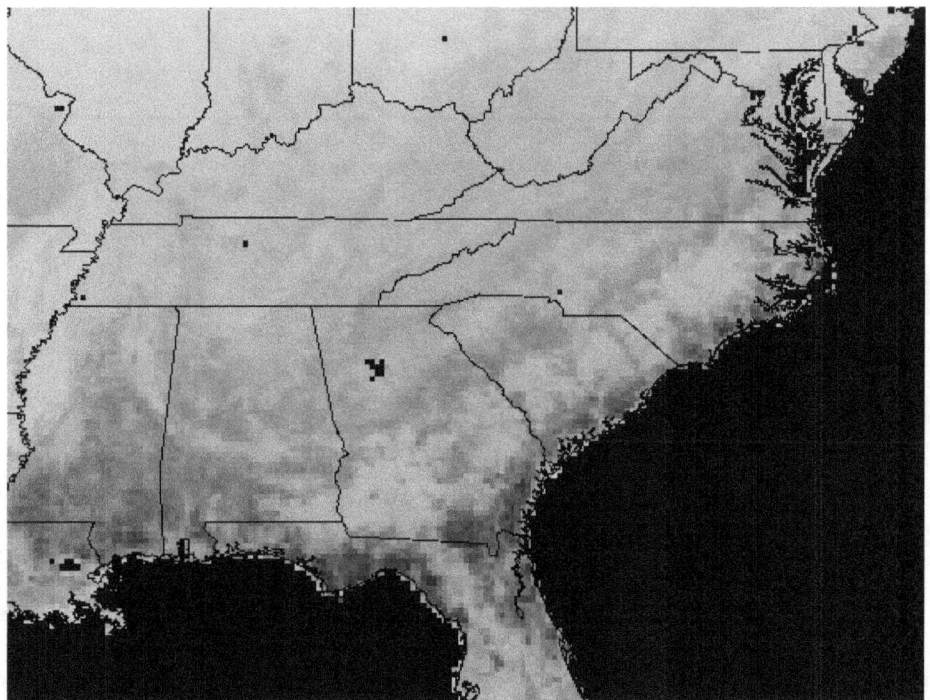

Figure 6. NASA EOS program MODIS instrument net primary vegetation (NPV) production for the southeast United States (1504031rgb-167772161.0). Darker = Greater NPV.

3.2.3.2 National Land Cover Dataset 1992 (NLCD 92)

The NLCD is a US land cover classification product based primarily on 1992 Landsat Thematic Mapper (TM) data. Land cover characteristics data describe the nature of the land surface at a particular location (Figure 7). The staff at the USGS, National Center for Earth Resources Observation and Science (EROS), in cooperation with the University of Nebraska-Lincoln, and the European Commission's Joint Research Centre compile land cover data as part of the Global Land Cover Characterization Program (GLCC). This effort is part of the NASA Earth System Science Pathfinder Program. The land cover information is drawn from Advanced Very High Resolution Radiometer (AVHRR) data and results in a 1-km resolution.

3.2.3.3 Frost free days/ hardiness zones

The number of frost free days defines the growing season available to many plants, particularly agricultural plants. A frost free map is closely associated with the USDA Hardiness Zone map used to determine the likely extent of particular plants and when new seedlings can be safely planted outside (Figure 8). Further it can be generated from the type of data that climate models output.

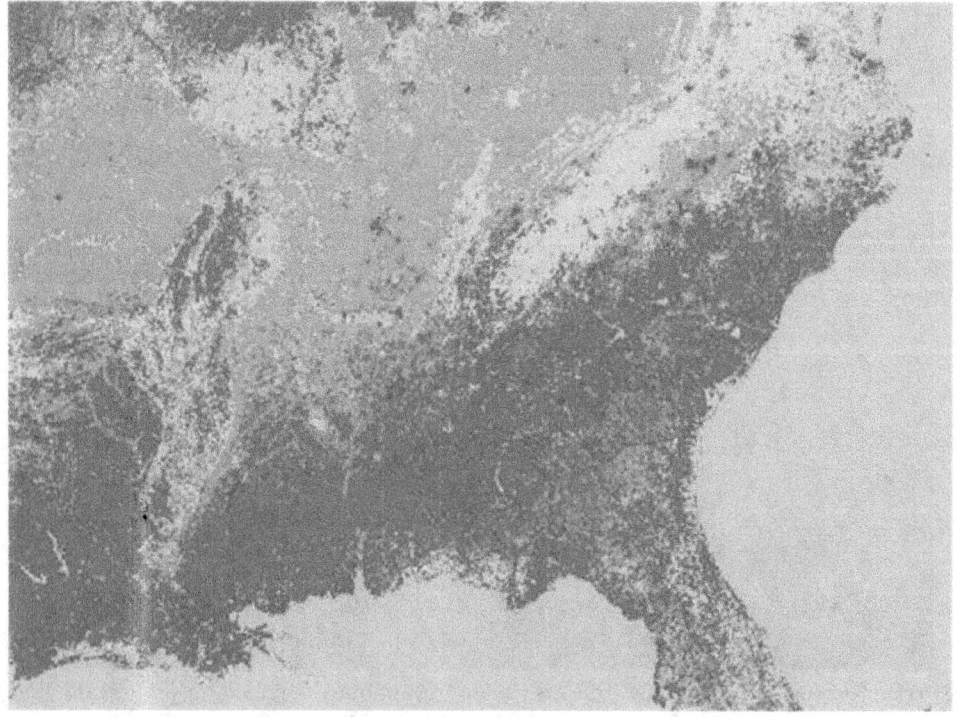

Figure 7. Land cover type derived from 1992 to 1993 1-km AVHRR data.

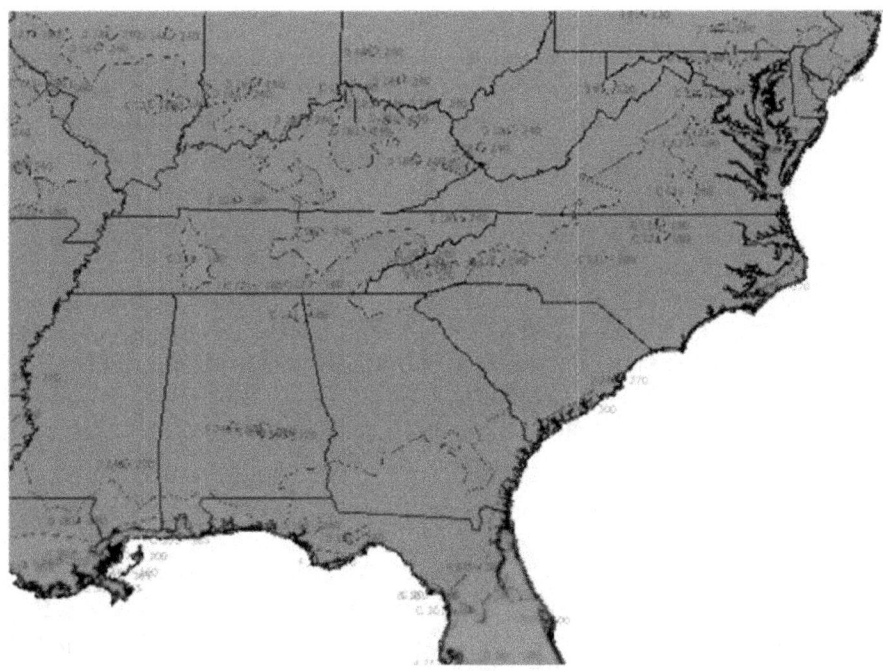

Source: http://cdo.ncdc.noaa.gov/climaps/

Figure 8. TMP07A13 median/mean length of freeze-free period (annual).

3.3 Reference for historical ecological characterizations

3.3.1 For Bailey's ecoregions

Bailey, Robert G. Identifying Ecoregion Boundaries. 2004. Environmental Management 34 (Supplement 1):S14-S26.
Summarizes the rationale used in identifying ecoregion boundaries on maps of the United States, North America, and the world's continents, published from 1976 to 1998. The geographic reasoning used in drawing boundaries involves 20 principles, which are presented to stimulate discussion and further understanding.

————. 1995. Description of the ecoregions of the United States (2d ed.). Misc. Pub. No. 1391, Map scale 1:7,500,000. Washington, DC: US Department of Agriculture (USDA) Forest Service.
This publication briefly describes and illustrates the Nation's ecosystem regions as shown on the map Ecoregions of the United States. A copy of this map is provided with the publication. The description of each region includes a discussion of land-surface form, climate, vegetation, soils, and fauna.

————. 1991. Design of Ecological Networks for Monitoring Global Change. Environmental Conservation 18(2):173-175.
World-wide monitoring of agricultural and other natural-resource ecosystems is needed in assessing the effects of possible climate changes and/or air pollution on our global resource-base. Monitoring of all sites is neither possible nor desirable for large areas, and so a means of choice has to be devised and implemented.

————. 1988. Ecogeographic Analysis. USDA Forest Service: Misc Publication 1465. Ecological units of different sizes for predictive modeling of resource productivity and ecological response to management need to be identified and mapped. A set of criteria for subdividing a landscape into ecosystem units of different sizes is presented, based on differences in factors important in differentiating ecosystems at varying scales in a hierarchy. Practical applications of such units are discussed.

————. 1984. The Factor of Scale in Ecosystem Mapping. 1984. Environmental Management 9(4):271-276.
Ecosystems come in many scales or relative sizes. The relationships between smaller and larger scales must be examined to predict the effects of management prescriptions on resource outputs. Environmental factors important in controlling ecosystem size change in nature with the scale of observation. Environmental factors that are thought to be useful in recognizing and mapping ecosystems at various scales are reviewed.

————. Testing an Ecosystem Regionalization. Journal of Environmental Management 19:239-248.
As a means of developing reliable estimates of ecosystem productivity, landscapes need to be stratified into homogeneous geographic regions. Such ecosystem regions are hypothesized to be productively different in important ways. One measure of the difference is hydrologic productivity. Data from 53 hydrologic bench-mark stations within major ecosystem regions were subjected to discriminate analysis. The ecosystem regions tested in this study exhibit a high degree of ability to circumscribe stations with similar hydrologic productivity.

————. 1983. Delineation of Ecosystem Regions. Environmental Management 7(4):365-373.
As a means of developing reliable estimates of ecosystem productivity, ecosystem classification needs to be placed within a geographical framework of regions or zones. This paper explains the basis for the regions delineated on the 1976 map *Ecoregions of the United States*. Four ecological levels are discussed — domain, division, province, and section — based on climatic and vegetational criteria. Statistical tests are needed to verify and refine map units

3.3.2 For Omernik's ecoregions

Omernik, J. M. 1987. Ecoregions of the conterminous United States (map supplement): Annals of the Association of American Geographers, 77(1):118-125, http://www.jstor.org/stable/2569206?cookieSet=1

Omernik, James M., and Robert G. Bailey. 1997. distinguishing between watersheds and ecoregions. Journal of the American Water Resources Association. 33(5):935-949.
Many state and Federal agencies have begun using watershed or ecoregion frameworks. Misunderstanding of each of the frameworks has resulted in inconsistency in their use and ultimate effectiveness. The focus of this paper is on the clarification of both frameworks. The issue is not whether to use watersheds or ecoregions frameworks, but how to correctly use the frameworks together.

4   Climate Change Modeling Review

4.1   General background to climate modeling

Climate Change as an area of concern dates back to the 1960s (Manabe and Wetherald 1967). Many individuals and groups have been working to objectively understand the direction of climate change and many models have been developed.* The best respected models all generate predictions based on a set of conventions disseminated through the Intergovernmental Panel on Climate Change (IPCC). Such standardization is meant to facilitate comparison between models. As the predictive capabilities of climate models are refined, discrepancies between them grow less significant. However, enough variation still exists that critics are able to use differences between the models to exaggerate the differences within climate research. To minimize such confusion, the IPCC acts as a coordinating organization and its reports are intended to reflect the scientific consensus amongst the experts in the field. That consensus includes items that should no longer be considered controversial (IPCC 2007):

- Climate change is occurring.
- Variations in temperature and precipitation occur locally.
- Globally, the planet earth is warming.

4.2   The scenarios on which climate modeling efforts are based†

One of the primary responsibilities of the IPCC is the arrangement of a series of standard future scenarios to assist with coordination and comparison between modeling results. This international standard set of scenario types is named after *The Special Report on Emissions Scenarios (SRES)*.

The SRES was prepared by the IPCC for the Third Assessment Report (TAR) in 2001 on future emission scenarios to be used for driving GCMs to develop climate change scenarios. The SRES were also used for the Fourth Assessment Report (AR4) in 2007. Four scenario families exist (Table 2).

---

* Some of the best known of which include National Center for Atmospheric Research (NCAR, in Boulder, CO, USA), the Geophysical Fluid Dynamics Laboratory (GFDL, in Princeton, NJ, USA), the Hadley Centre for Climate Prediction and Research (in Exeter, UK), the Max Planck Institute for Meteorology in Hamburg, Germany, and the Institut Pierre-Simon Laplace (IPSL in Paris, France).

† This section draws heavily from: IPCC (2007a).

Table 2. The four *SRES* scenario families of the *Fourth Assessment Report* with associated projected global average surface temperature increase by 2099

Homogenous: Global*	A1 rapid economic growth (includes groups: A1T; A1B; A1FI) +1.4 – 6.4 °C	B1 global environmental sustainability +1.1 – 2.9 °C
Heterogeneous: Regional / Local	A2 regionally oriented economic growth +2.0 – 5.4 °C	B2 local environmental sustainability +1.4 – 3.8 °C

* Table format drawn partially from: http://www.grida.no/publications/other/ipcc_tar/?src=/climate/ipcc_tar/wg3/081.htm

The A1 scenarios describe "a future world of very rapid economic growth, global population that peaks in mid-century and declines thereafter, and the rapid introduction of new and more efficient technologies." (IPCC 2007a, p 18) Major underlying themes of the A1 family of scenarios are:

• convergence among regions
• capacity building and increased cultural and social interactions
• substantial reduction in regional difference in per capita income.

The three A1 groups are distinguished by their technological emphasis:

• A1F1 – Fossil Intensive
• A1T— Non-fossil energy sources
• A1B – A balance across all energy sources.

The A2 scenarios describe a very heterogeneous world. Major underlying themes of the A1 family of scenarios are:

• self reliance
• preservation of local identities
• continuously increasing population
• regionally oriented economic development
• more fragmented and slower per capita income growth and technological change.

The B1 scenarios describe a convergent world with the same global population that peaks in mid-century and declines thereafter (just as in A1), but with rapid change in economic structures toward a service and information economy. Major underlying themes of the A1 family of scenarios are:

• reductions in material intensity
• introduction of clean and resource efficient technologies
• emphasis on global solutions to economic, social and environmental sustainability.

The B2 scenarios describe a world in which is oriented toward environmental protection and social equality, but focuses on local solutions to economic, social and environmental sustainability. Major underlying themes of the A1 family of scenarios are:

- continuously increasing global population (at a lower rate than A2)
- intermediate levels of economic development
- less rapid and more diverse technological change than in B1 and A1 scenarios.

4.3 The major climate models

Since the 1990s, the international climate change science community has participated in a series of efforts (often called *campaigns*) to carry out major, mostly coordinated attempts to exercise their best available modeling capabilities under similar sets of SRES scenarios. The most recent is termed "AR4." In this study, we used AR4 model results in our analysis. Table 3 lists the major players in the AR4 campaign and the status of their models (IPCC 2010). The next major coordinated modeling effort will be the IPCC Fifth Assessment Report (AR5), which will be finalized in 2014 – modeling efforts for AR5 have already begun.

4.4 Which predictive models to choose for this work?

Of all the available models, which are "best" for future predictions of different variables? One objective criterion would be the models that have the closest validation with the variables of interest here, for example Precipitation over CONUS compared to some climatology. Those models having had the greatest number of validation studies and those with the longest-period of development (one to two decades) include:

- CM2.1 (GFDL model -NOAA Princeton)
- E-H and E-R (NASA GISS)
- HadGEM1 (UKMO)
- CGCM3 (Canadian [CCCma] model)
- CCSM3 (National Center for Atmospheric Research [NCAR] Boulder).

Other models have shorter lifetimes of development since inception, and less person-hours involved in validation and calibration. Table 4 lists the expected temperature increases among some of these models (IPCC 2001). Those used in this report are shaded.

Table 3. SRES scenario runs for AR4 (status of data: August 2006).

Center	Country	Acronym	Model
Beijing Climate Center	China	BCC	CM1
Bjerknes Centre for Climate Research	Norway	BCCR	BCM2.0
Canadian Center for Climate Modelling and Analysis	Canada	CCCma	CGCM3 (T47 resolution) CGCM3 (T63 resolution)
Centre National de Recherches Meteorologiques	France	CNRM	CM3
SHADE (Australia's Commonwealth Scientific and Industrial Research Organization)	Australia	CSIRO*	Mk3.0
Max-Planck-Institut for Meteorology	Germany	MPI-M	ECHAM5-OM
Meteorological Institute, University of Bonn, Germany		MIUB	ECHO-G
Meteorological Research Institute of KMA, Korea		METRI	
Model and Data Groupe at MPI-M, Germany		M&D	
Institute of Atmospheric Physics	China	LASG	FGOALS-g1.0
Geophysical Fluid Dynamics Laboratory	USA	GFDL	CM2.0 CM2.1
Goddard Institute for Space Studies	USA	GISS	AOM E-H E-R
Institute for Numerical Mathematics	Russia	INM	CM3.0
Institut Pierre Simon Laplace	France	IPSL	CM4
National Institute for Environmental Studies	Japan	NIES	MIROC3.2 hires MIROC3.2 medres
Meteorological Research Institute	Japan	MRI	CGCM2.3.2
SHADE National Centre for Atmospheric Research	USA	NCAR	PCM** CCSM3
SHADE UK Met Office	UK	UKMO	HadCM3 HadGEM1
National Institute of Geophysics and Volcanology	Italy	INGV	SXG 2005

*The Commonwealth Scientific and Industrial Research Organisation (CSIRO), Australia's national science agency
**Parallel Climate Model

Table 4. Temperature increase 2000 to 2100 spread among selected GCMs.

Model	Total		Land		Ocean	
	°F	°C	F	°C	F	°C
CCSR/NIES	40.5	4.7	44.6	7.0	38.8	3.8
CCCma	39.2	4.0	41.0	5.0	38.5	3.6
CSIRO	38.8	3.8	40.8	4.9	38.1	3.4
Hadley Centre	-38.7	3.7	41.9	5.5	37.4	3.0
GFDL	37.9	3.3	39.6	4.2	37.4	3.0
MPI-M	37.4	3.0	40.3	4.6	36.3	2.4
NCAR PCM	36.1	2.3	37.6	3.1	35.6	2.0
NCAR CSM		2.2	36.9	2.7	35.6	2.0

Figure 9 shows a comparison of the 16 models for temperatures and precipitation from "bias-corrected" data (more about "bias-corrected" data later in section 4.5 this chapter) for the default study area of this research in the South East US quadrant. The purpose of the graphic is to highlight a few ideas that are evident when one compares different models using any of a large number of characteristics that could be investigated:

- Differences between models exist.
- Variations between models are in terms of a few percent, not orders of magnitude.
- Variations exist, but all models agree that warming will occur, i.e., not one generates a cooling trend.
- Similarities among the major models are more notable than differences.

We have chosen to use the CCCma CGCM3.1 (Canadian), CSIRO MK3 (Australian), Hadley HADCM3.1 (United Kingdom) and the NCAR PCM 1.1 (United States) models. These The Hadley and PCM models were chosen to represent relative extremes in GCM forecasts. The Canadian and Australian models were chosen to represent moderate forecasts. Hadley model will likely show the greatest predicted changes.

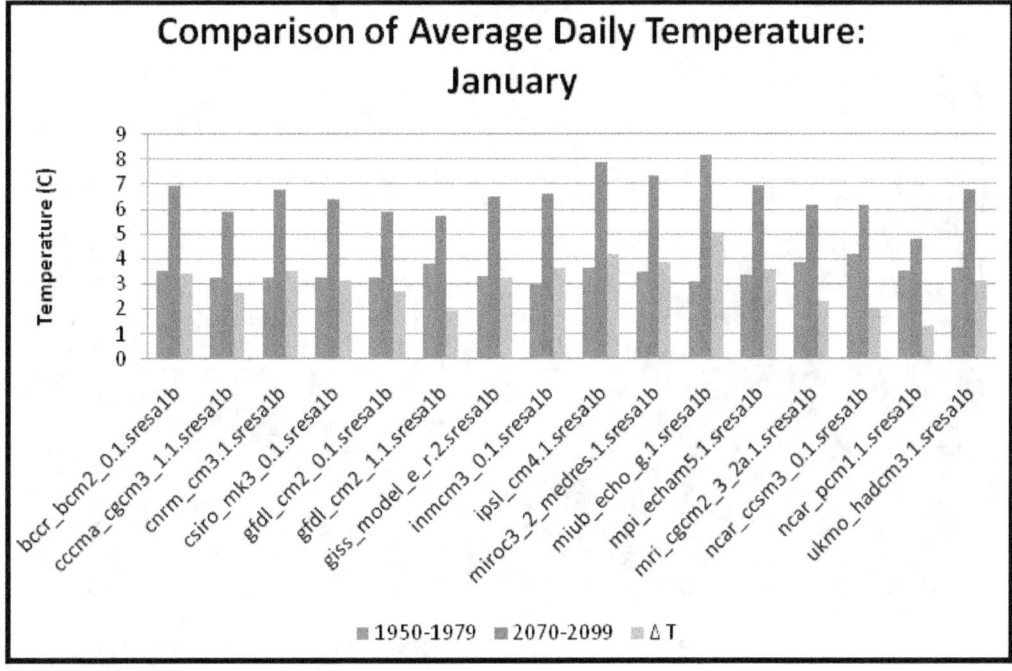

Figure 9. Comparison of the 16 models for temperature data for the south east US quadrant.

We wished to ensure that the models represented a range of nationalities so that this report could not be perceived as bias toward American or Military concerns. The Canadian and Australian models are the basis for the temperature, precipitation and ecosystem change analyses. The Hadley and PCM models are used as the basis for the Hargrove Ecosystem Change analysis. For this report we could easily have chosen any other combination of climate models, or preferably used all of them. The limitation of time required choices and we believe those that we have made well represent the range of modeling possibilities.

4.5   Climate models used for this work

4.5.1 Canadian global climate model (CGCM3)

The third version of the coupled Canadian global climate model (CGCM3) [Figure 10] makes use of the same ocean component as the earlier CGCM2, but of the mainly updated atmospheric component AGCM3 [third-generation atmospheric general circulation model]. The sea-ice component is a two-category model (mean thickness and concentration), except that a prognostic equation for ice concentration follows Hibler (1979).

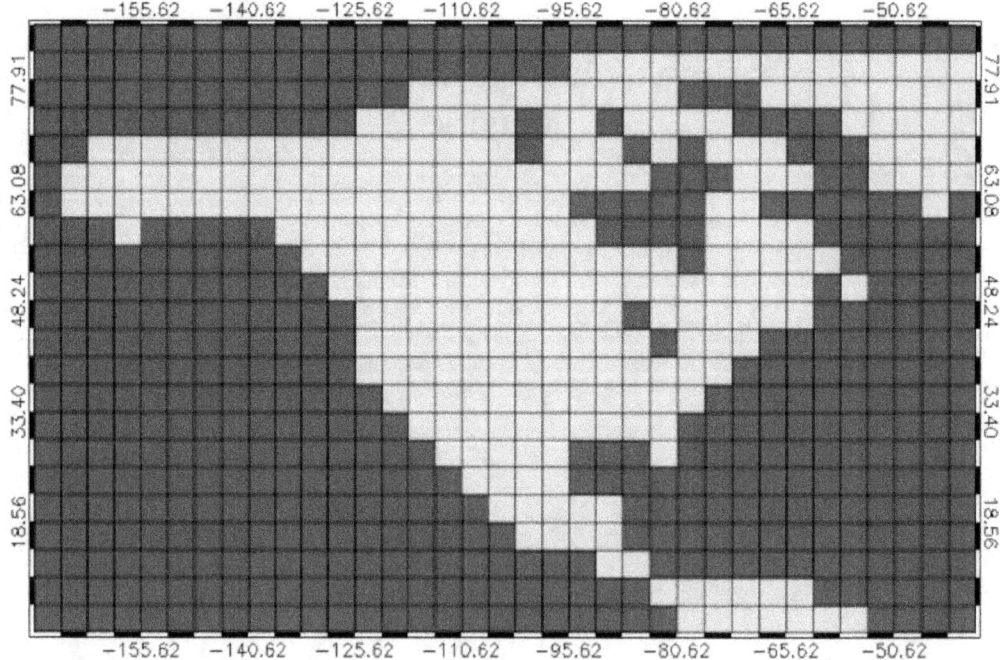

Figure 10. CGCM3 land/sea mask over North America. Land grid points are in green.

The spectral representation of T47 in AGCM3 comprises 47 wave triangularly truncated spherical harmonic expansion (which provides roughly 3.75° x 3.75° surface grid resolution), and includes 32 vertical levels, extended to 1hPa (e.g., ~ 50km above surface). Some new key features are included: the introduction of CLASS a new module for treatment of the land surface processes (Verseghy et al, 1992); the moist convective adjustment algorithm that was used in AGCM2 has been replaced by the cumulus parameterization of Zhang and McFarlane (1995); the quantity of transported water vapour is the hybrid moisture variable proposed by Boer (1995), which makes a significant difference between the second and third generation models (GEC3, Environment Canada, and DRI 2010).

4.5.2 References for CGCM3

Boer, G. J. 1995: A hybrid moisture variable suitable for spectral GCMs. Research Activities in Atmospheric and Oceanic Modelling. Report No. 21, WMO/TD-No. 665. World Meteorological Organization, Geneva, Switzerland.

Hibler, W. D., 1979. A dynamic thermodynamic sea ice model. J. Phys. Oceanogr. 9:815-846.

Verseghy, D. L., N. A. McFarlane, and M. Lazare. 1993. A Canadian Land Surface Scheme for GCMs:II. Vegetation model and coupled runs. Int. J. Climatol. 13:347-370.

Zhang, G. J. and N. A. McFarlane. 1995: Sensitivity of climate simulations to the parameterization of cumulus convection in the CCC-GCM. Atmos.-Ocean. 3:407-446.

4.5.3 British climate model (HadCM3)

HadCM3 is the third version of the coupled atmosphere-ocean model in Gordon et al. (2000). Unlike the Canadian model, it does not use surface flux adjustment procedures. The atmospheric component of the model uses 19 levels with a regular **horizontal resolution** of **2.5° in latitude x 3.75°** in longitude, thus consisting of a global 96 x 73 points grid. This corresponds to an approximate resolution of 417 x 278 km at the equator and 295 x 278 km at 45° of latitude (comparable to a T42 spectral resolution). The oceanic component of the model uses 20 vertical levels with a horizontal resolution of 1.25° x 1.25°. Contrary to CGCM2, the HadCM3 model systematically counts 360 days in 12 months of 30 days each. (GEC3, Environment Canada, and DRI 2010).

4.5.4 Reference for HadCM3

Gordon, Chris, Claire Cooper, Catherine A. Senior, Helene Banks, Jonathan M Gregory, Timothy C Johns, John F. B. Mitchell, and Richard A. Wood. 2000. The simulation of SST, sea ice extents and ocean heat transports in a version of the Hadley Centre Coupled Model without flux adjustments. Climate Dynamics. 16:147-168. Bracknell, UK: Hadley Centre for Climate Prediction and Research.

4.5.5 American parallel climate model (PCM1)

Sponsored by the Department of Energy, PCM1 is the first version of a joint effort to develop a parallel climate model between Los Alamos National Laboratory (LANL), the Naval Postgraduate School (NPG), the US Army Corps of Engineers' CRREL, and the National Center for Atmospheric Research (NCAR). We have coupled the NCAR Community Climate Model version 3 (CCM3), the LANL Parallel Ocean Program, and an NPG sea ice model together in a massively parallel computer environment.

Based on the experience with the NCAR Climate System Model, to minimize the initial drift of the coupled system, the ocean/ice can be spun-up with forcing from previous CCM3 runs with prescribed ocean temperatures. This has also been useful in demonstrating and improving the kind of adjustments that occur in the ocean and ice due to coupling the CCM3, without having to run the more expensive coupled system. The full system has been in full production with several control experiments and many ensemble climate change simulations in progress and completed (Figure 11).

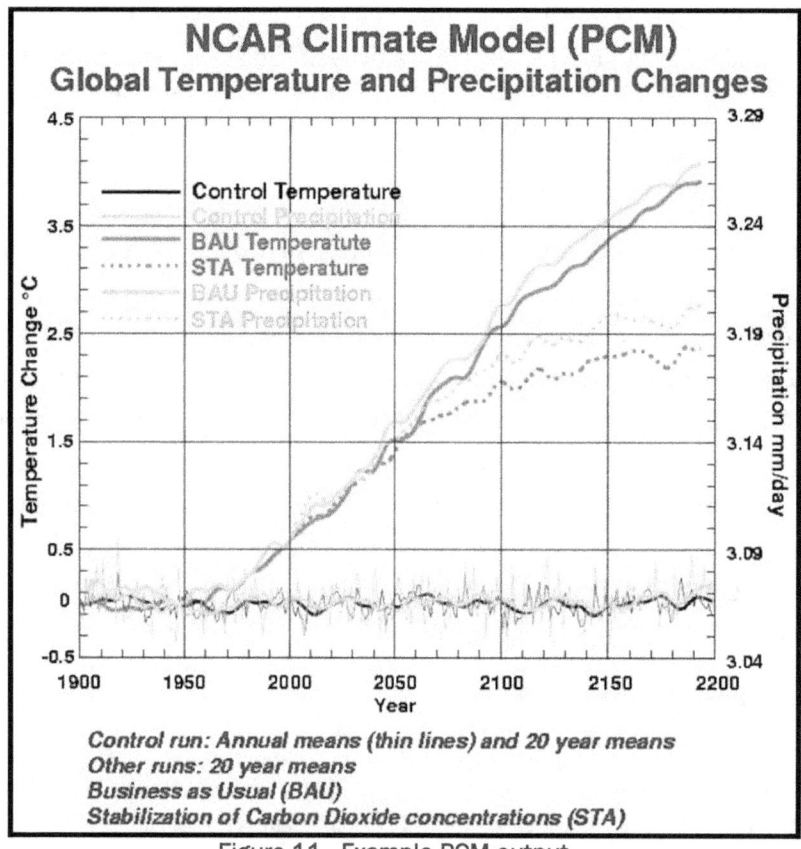

Figure 11. Example PCM output.

4.5.6 References for PCM1

Intergovernmental Panel on Climate Change (IPCC). 2007. S. Solomon, D. Qin, M. Manning, Z. Chen, M. Marquis, K.B. Averyt, M. Tignor and H.L. Miller (eds.). 2007. *The Physical Science Basis, Contribution of Working Group I to the Fourth Assessment Report of the Intergovernmental Panel on Climate Change.* Cambridge, United Kingdom and New York, NY, USA: Cambridge University Press, http://ipcc-wg1.ucar.edu/wg1/wg1-report.html

Weatherly, J.W., and C. M. Bitz. 2001. Natural and anthropogenic climate change in the arctic. 12th Symposium on Global Change and Climate Variations, 15-18 January 2001, Albuquerque, Boston, MA: American Meteorological Society.

4.5.7 Australian climate model (CSIRO MK3)

The CSIRO model Mk3.5:

> represents a significant improvement over the preceding Mk3.0 model in aspects relevant to oceanic regions near Australia. The CSIRO Mk3.5 model will be an important tool for climate modelers studying the impacts of climate change, for example in simulating rainfall changes over southern Australia. The most significant improvements result from the use of a more physically realistic set of parameters to represent the transport of heat and freshwater by oceanic eddies. It also features considerably more realistic circulation and stratification in the Southern Ocean. (CSIRO 2006). (Figure 12 shows an example of CSIRO model output.)

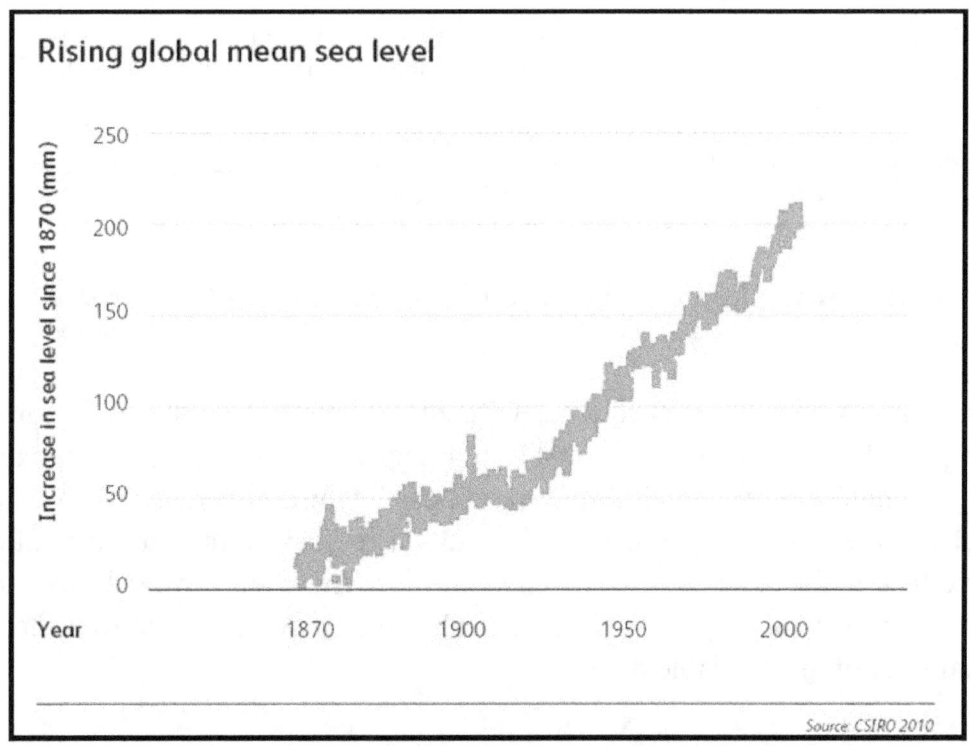

Figure 12. Rising global mean sea level based on the CSIRO model.

4.5.8 References for CSIRO MK3

Cai, W., M. A. Collier, P. D. Durack, H. B. Gordon, A. C. Hirst, S. P. O'Farrell, and P. H. Whetton. 2003. The response of climate variability and mean state to climate change: Preliminary results from the CSIRO Mark 3 coupled model. CLIVAR Exchanges. 28:8-11.

Cai, W., G. Meyers, and G. Shi. 2005. Transmission of ENSO signals to the Indian Ocean. Geophys. Res. Let. In press.

Cai, W., G. Shi, Y. Li. 2005. Multidecadal fluctuations of winter rainfall over southwest Western Australia simulated in the CSIRO Mark 3 coupled model. Geophys. Res. Let. submitted.

Cai, W., H. Hendon, and G. Meyers. 2005. An Indian Ocean Diploe-like variability in the CSIRO Mark 3 climate model. J. Climate. In press.

Cai, W., M. A. Collier, H. B. Gordon, and L. J. Waterman. 2003. Strong ENSO variability and a super-ENSO pair in the CSIRO coupled climate model. Monthly Weather Review. 131:1189-1210.

Cai, W., M. J. McPhaden, M. A. Collier. 2004. Multidecadal fluctuations in the relationship between equatorial Pacific heat content anomalies and ENSO amplitude. Geophys. Res. Let. 31, L01201. DOI:10.1029/2003GL018714.

Gordon, H. B., L. D. Rotstayn, J. L. McGregor, M. R. Dix, E. A. Kowalczyk, S. P. O'Farrell, L. J. Waterman, A. C. Hirst, S. G. Wilson, M. A. Collier, I. G. Watterson, and T. I. Elliott. 2002. The CSIRO Mk3 climate system model [Electronic publication]. CSIRO Atmospheric Research technical paper No. 60. Aspendale: CSIRO Atmospheric Research, http://www.dar.csiro.au/publications/gordon_2002a.pdf

Watterson, I. G., 2005: The intensity of precipitation during extra-tropical cyclones in global warming simulations: a link to cyclone intensities? Tellus A. Accepted for publication.

Watterson, I. G., and M. R. Dix. 2005. Effective sensitivity and heat capacity in the response of climate models to greenhouse gas and aerosol forcings. Q. J. Roy. Met. Soc., 131:259-280.

4.6 Problems with the output of the climate models used for this work

There are two important problems with climate models. First, the major outputs that would be useful for the characterization of ecological changes are temperature and precipitation (and possibly humidity). Although other outputs are available (Figure 13), they do not relate well to issues that have an effect on the ecosystem's existence. Because the climate models are the predictive vehicle for this work, this is a limitation on our work from the character of the available data.

OUTPUT	Grid, resolution, frequency of output	INPUT	Grid, resolution, frequency of input
Humidity : Specific (SHUM) and Relative (RHUM)	Gaussian grid (97x48 pts)	Model humidity (ES), Temperature (TEMP) and Surface Pressure (LNSP)	Spectral grid T47
	Standard pressure levels		ETA15 vert levels
	daily		6 hours
Geopotential (PHI)	Gaussian grid (97x48 pts)	Orography (PHIS), Temperature (TEMP), Surface Pressure (LNSP), and RGASM²	Spectral grid T47
	Standard pressure levels		ETA15 vert levels
	daily		6 hours
Mean Sea Level Pressure (PNM)	Gaussian grid (97x48 pts)	Orography (PHIS), Temperature (TEMP), and Surface Pressure (LNSP)	Spectral grid T47
	Standard pressure levels		-
	daily		6 hours
Temperature (TEMP)	Gaussian grid (97x48 pts)	Temperature (TEMP) [and for gsaplt : Orography (PHIS), and Surface Pressure (LNSP)]	Spectral grid T47
	Standard pressure levels		ETA15 vert levels
	daily		6 hours
Temperature at 2m (ST)	Gaussian grid (97x48 pts)	Temperature at 2m (ST)	Gaussian grid (97x48 pts)
	surface		surface
	daily		daily
	K		℃
Accumulated Precipitation (PCP)	Gaussian grid (97x48 pts)	Accumulated Precipitation (PCP)	Gaussian grid (97x48 pts)
	surface		surface
	daily		daily
Specific humidity at 2m (SQ)	Gaussian grid (97x48 pts)	Specific humidity at 2m (SQ)	Gaussian grid (97x48 pts)
	surface		surface
	daily		daily

Figure 13. List of CGCM3.1 T47 variables directly available and those derived/interpolated.

The second issue is potentially even more problematic — the geographically referenced data is gross in size. In fact, most GCMs output their results in a grid format that is roughly 3 degrees by 3 degrees in size (Figure 10). To get a better impression of the scale of GCM output, refer to the green grid cell in Figure 14. We felt that using such generalized data would probably result in less than satisfactory results.

4.7 Better data resolution using the bias corrected and downscaled World Climate Research Programme (WCRP) CMIP3 climate projections

Assessing the risks for climate impacts at the local and regional scales necessary for ecological modeling requires "downscaling," or refining, the climate model results to a finer spatial grid resolution than the 300-500 km grid used in GCMs. Ideally the downscaling could also correct or adjust the GCM outputs for the biases between the coarse-scale GCMs and local-scale features in temperatures, precipitation and associated variables, which are created by local-scale topography, landscape types, vegetation, and water bodies.

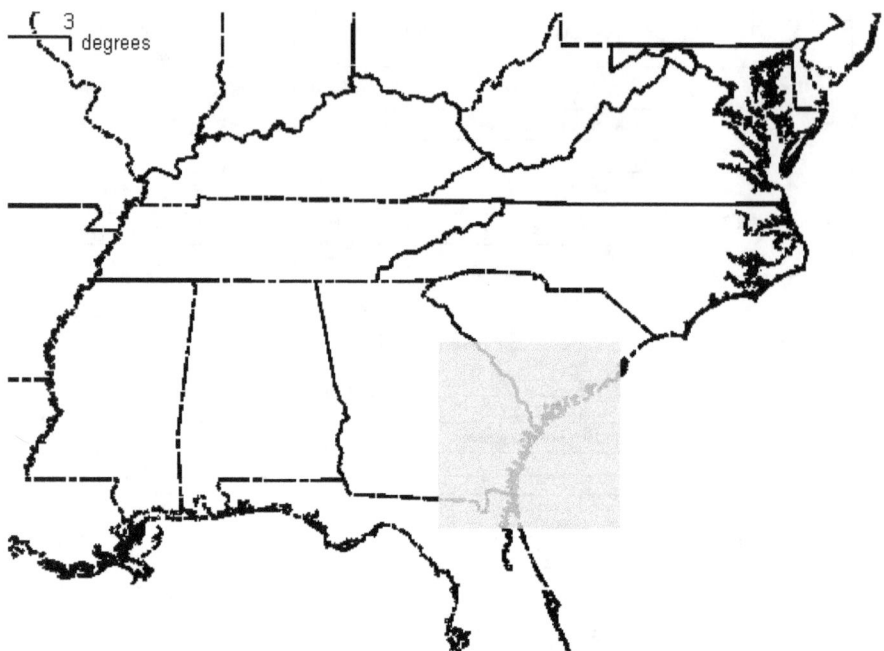

Figure 14. The coverage of a single temperature data point from Canadian CGCM3 model is shown in green.

The two primary methods used for downscaling are dynamical (using a dynamic numerical weather model or regional climate model to simulate temperature, moisture, and winds and using GCM output as boundary conditions), and statistical (using observational weather and local topography data to make adjustments to GCM outputs on a local scale). The statistical downscaling approach has the advantage of using bias correction for adjusting the GCM outputs to agree with the local climate observations in locations where sufficient local observations are available. The statistical approach is also less computationally intensive than dynamic modeling.

This current report uses the results of the bias-corrected spatial downscaled (BCSD) GCM projections of future climate scenarios produced by Maurer et al. (2007), which are the GCMs archived by the WCRP's Coupled Model Intercomparison Project phase 3 (CMIP3) multi-model dataset (see Meehl et al. 2007). The archived GCM outputs include those described in sections 4.1 and 4.2 above. The BCSD approach used by Maurer et al. (2007) is described in the references, and on their Program for Climate Model Diagnosis and Intercomparison (PCMDI) website.[*]

---

[*] http://gdo-dcp.ucllnl.org/downscaled_cmip3_projections/dcpInterface.html#About

The approach can be summarized (from their description) by two steps:

1. *Bias correction.* To correct for the differences between each GCM's simulation and the observed data for surface-level temperature and precipitation using a quantile mapping technique. Observed gridded climate data from 1950-1999 is compared to the GCMs simulated variables of this period on a 2-degree grid cell-by-cell basis, and cumulative distribution functions (CDF) for each are generated. These CDFs are used to bias-correct the GCM variables for the 1950-1999 period to the mean and variance as the observed. The same GCM's simulation of the 21st century is then similarly adjusted to include the trend from the GCM between the two periods and the adjusted variance from the observations. This assumes that the spatial and temporal variance in the 21st century is unchanged, which is a limitation of this technique.

2. *Spatial downscaling.* With the bias correction completed on the 2-degree grid (approx 200 km), the conversion to the 1/8-degree grid (12 km) is performed essentially by a modified inverse-distance-squared interpolation on a time-by-timestep basis. There are additional adjustment factors for temperature and precipitation (GCM/Observed) that ensure the spatial patterns of each are preserved in this interpolation.

The BCSD dataset includes 112 climate projections over the contiguous United States from CMIP3, including 16 GCMs from research institutions worldwide, with three IPCC SRES scenarios: A1b (blended growth in fossil and non-fossil fuel) A2 (slower growth), and B1 (slower growth, stabilization). The analysis in the following section uses the bias-corrected downscaled 1/8-degree data from the four GCMs shaded in Table 4, using the mean temperature and precipitation for three selected years, 2000, 2050, and 2099, and for two scenarios, A1b and B1). The two scenarios are a set of "bookends" on the more likely (or possible) future growth of emissions in the 21st century. The A1b emission scenario generally produces the more rapid changes in temperature than the B1 scenario, which has slower growth and stabilization of CO_2 concentrations after 2050.

Figure 15 shows an example of the results drawn from BCSD data for the GCMs. The 16 GCMs show precipitation rates for the southeast United States between 3.0 and 3.5 mm/day for 1950-1979. The observed rate for Fort Stewart in the first column is 2.5 mm/day, reflecting the typical regional difference for this site from the large-scale average.

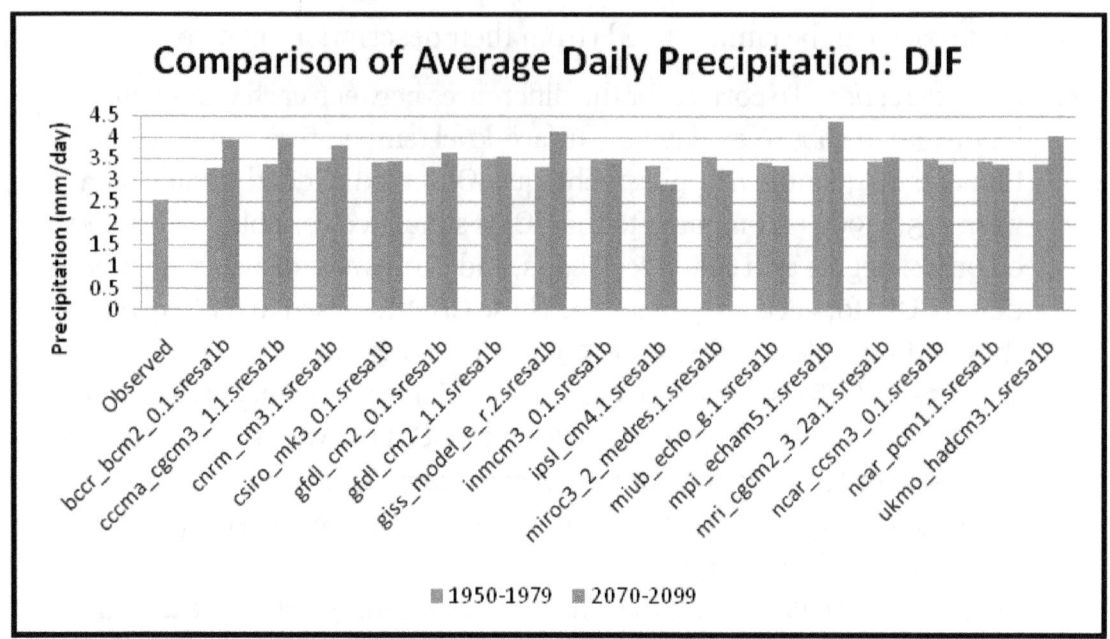

Figure 15. Comparison of average precipitation rates (mm/day) for Dec.-Jan.-Feb. for the southeast United States from the downscaled data for 16 GCMs for the 1950-1979 and projected 2070-2099 period (A1b scenario), and the observed precipitation rate from Fort Stewart site data for 1950-1979.

For the 2070-2099 period in the projected A1b scenario, the precipitation rates vary from 2.9 to 4.4 mm/day, and 10 of the 16 GCMs exhibit increases. This variance of changes among the GCMs is typical for the precipitation over the United States, where the mid-latitudes lie between zones of less precipitation to the south and greater precipitation to the north. By contrast, the mean temperature responses for A1b for this region (Figure 8, p 19) are all positive, 2 to 5 °C (3.6 and 9 °F).

Other components of the parent project ERDC Framework for Assessing the Environmental Effects of Climate Change for the Military are developing dynamical modeling capabilities to produce climate variables from GCM scenarios at finer resolution than 12 km (Weatherly 2011). The dynamical model is responsive to local-scale terrain, and can incorporate the feedback of changes in surface characteristics of wetlands, vegetation, or drought on the local climate impacts. For this present survey-scale study across multiple installations, the statistical downscale approach is sufficient to consider the changes in mean temperature and precipitation among a finite number of GCMs and scenarios from the CMIP3 archive.

4.8 Drawbacks and limitations to downscaled climate model projection data

There are several key limitations to performing and using both statistical and dynamic downscaling — some are scientific misconceptions, others are technical limitations. One misconception is that the downscaled GCM climate projections on a finer grid resolution may represent a more accurate prediction of the future large-scale temperatures. If the large-scale GCM projections exhibit a temperature increase of 2 °C (3.6 °F) over 80 years over a region of the United States, that change will be reflected in the downscaled data on the finer scale, with local variations associated with changes in elevation and surface types. The actual future temperature increase in 80 years may turn out to be 5 °C (9 °F) for that region — the downscaling process would not improve this predictive skill. The reasons for the large-scale discrepancy could be any number of causes: limitations in the GCMs' physics in atmosphere, ocean, or terrestrial components, or differences between actual CO_2 concentrations observed in the future and those assumed for the future GCM simulations.

The technical limitations to dynamical model downscaling are also potentially numerous as the approach is based on another complex numerical meteorological dynamical model with feedbacks involving terrestrial processes, vegetation, clouds, and radiation. These regional feedbacks are useful for refining the processes that BCSD approach cannot (i.e., the changes in terrestrial and hydrologic processes at higher time and spatial resolutions). Because of the significant computational time and resources needed, it is not practical to complete multiple high-resolution regional-model runs for each CO_2 emission scenario (A1, B1, etc) for every sub-region of interest. Owing to this limitation, the North American Regional Climate Change Assessment Program (NARCCAP; Mearns et al. 2009) is archiving a subset of the matrix of six GCMs and four RCMs for its intercomparison, using the current NCEP climatology and one IPCC A2 emission scenario for most of the North American continent.

The technical limitations to dynamical model downscaling are also potentially numerous: as the approach is based on another complex numerical meteorological dynamical model with feedbacks involving terrestrial processes, vegetation, clouds and radiation. These regional feedbacks *are* useful for refining the processes that BCSD approach cannot, i.e., the changes in terrestrial and hydrologic processes at higher time and spatial resolutions. Because of the significant computational time and resources needed to complete multiple high-resolution regional-model runs, it is not practical CO_2 emission scenario (A1, B1, etc), or for every sub-region of interest for different investigations. Owing to this limitation, the North American Regional Climate Change Assessment Program (NARCCAP, *Mearns et al.* 2009) is archiving a subset of the matrix of 6 GCMs and 4 RCMs for its intercomparison, using the current NCEP climatology and one IPCC A2 emission scenario.

The NARCCAP approach also illustrates the most frequently criticized aspect of a common downscaling approach – using GCM output variable *directly* as boundary conditions for regional models. As many critics point out, this approach simply propagates the errors and biases across the different GCMs and has little value in improving the *accuracy* of these models in relation to the observed climate or towards predicting future climate change. A more justified approach would be analogous to that taken by the BCSD, by using observed climate and weather data in combination with the trends and changes simulated by the GCM-generated climate scenarios.

For the ongoing ERDC dynamical downscaling modeling project, this modified approach with observed climate data and BCSD output is underway (Weatherly, 2011). To drive the regional climate model, the observed climate data are used first as the boundary conditions for the control case, and BCSD temperatures are added to the observed data for future climate cases.

Maurer, E. P., L. Brekke, T. Pruitt, and P. B. Duffy. 2007. Fine-resolution climate projections enhance regional climate change impact studies. *Eos Trans. AGU.* 88(47):504.

Mearns, L. O., W. J. Gutowski, R. Jones, L.-Y. Leung, S. McGinnis, A. M. B. Nunes, and Y. Qian. 2009. A regional climate change assessment program for North America. *Eos Trans. AGU.* 90(36):311-312.

Meehl, G. A., C. Covey, T. Delworth, M. Latif, B. McAvaney, J. F. B. Mitchell, R. J. Stouffer, and K. E. Taylor. 2007. The WCRP CMIP3 multi-model dataset: A new era in climate change research, *Bulletin of the American Meteorological Society.* 88:1383-1394.

5   Description of Analysis Procedures to Generate Predictive Map Data Sets

5.1   Input 1 — Ecological characterization of the "current" situation

In this section, we determined to focus on ecological characterization data sets, which the literature search suggested were of the greatest interest to our investigation. These became the basis of the "current" situation. Using the characteristics of the changes as determined from the "predictive models" (below), the purpose was to modify the current categorization to reflect the character of the future ecosystem spatial distribution.

Bailey's Ecoregions dataset was considered pertinent because it is a classic representation of ecosystem distribution that is widely recognized as a standard. It would be a mistake to carry out an investigation like this one without including Bailey's work.

Omernik's Ecosystems map is another standard ecosystem product. It is map with which most individuals in the discipline are also familiar. Changes in this map compared to changes in the Baileys map will tend to make clear and reemphasize how climate change has an effect on the population in general as well as the military.

GAP is not as well known as *Bailey's* or *Omernik's*; it was considered pertinent because it was generated through a large cooperative effort among many state and regional agencies over several years and has been refined to much greater detail than can be done by an individual. In particular more local groups and individuals have a "buy-in" to this delineation because they are largely the sources of the information needed to generate it. The degree of detail is in the same range as the downscaled climate data used in conjunction with it so there exists a "good fit" among data types.

The *Hargrove/Hoffman Potential of Multivariate Quantitative Methods for Delineation and Visualization of Ecoregions* was considered pertinent because it was originally targeted at ecosystem change issues, is predictive in the correct time horizon for this project, and provides an independent evaluation technique compared to the others used in this work.

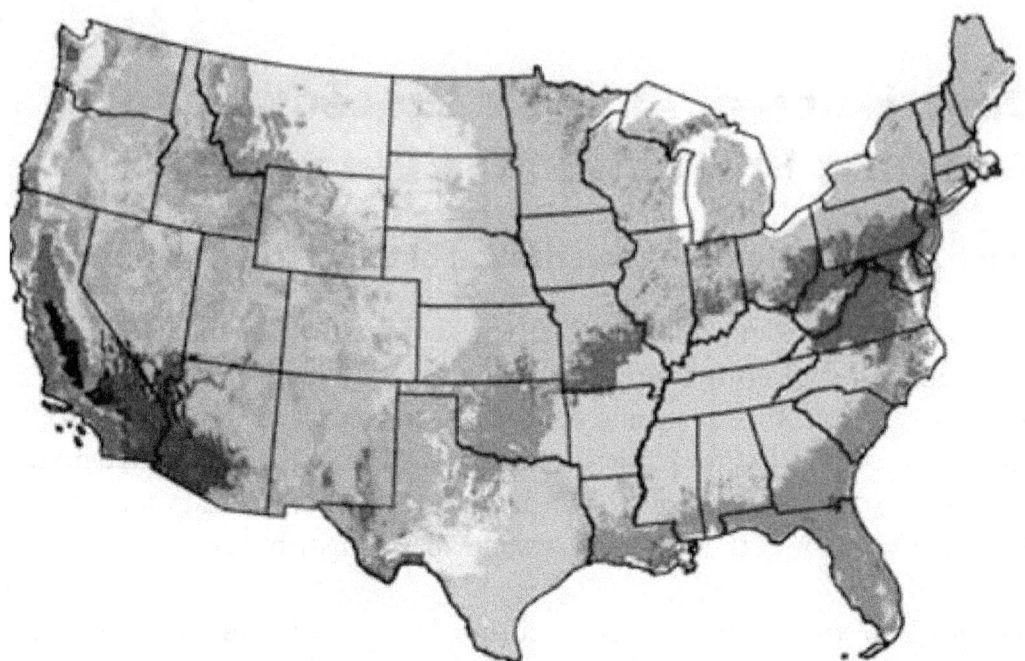

Figure 16. An initial Hargrove/Hoffman map of the United States. One hundred common ecoregions were found representing predictions for the year 2099 from the Hadley global climate simulations. In this technical report, we used more recent data covering the globe with 30,000 non–traditional "ecoregions."

Similarity of any ecoregion to all other ecoregions can be quantified and displayed as a "representativeness" map (Figure 16). The data set is particularly interesting because it is generated in a highly objective manner with the latest technology and characterizes ecosystems in a fundamental physical science manner.

5.2 Input 2 — Climate change data used as the basis of the "future" situation

Our work is based on standard climate models. Of the many that are available, we choose to use four standard models for the characterization of the 2100 state of affairs. Further, we adopted the following ecosystem-climate-scenario combinations:

1. Bailey's Ecoregions using
   a. *Canadian GCM3.1 model*
      (1) *Scenario A1b Globalized Rapid Economic Growth*
      (2) Scenario B1 Globalized Environmental Sustainability
   b. Australian Mk3.0 model, Scenario A1b Globalized Rapid Economic Growth

2. Omernik's Ecoregions using
   a. Canadian GCM3.1 model
      (1) Scenario A1b Globalized Rapid Economic Growth
      (2) Scenario B1 Globalized Environmental Sustainability
   b. Australian Model, Scenario A1b Globalized Rapid Economic Growth
3. GAP Analysis based on the Hargrove/Hoffman Ecoregions using
   a. Hadcm3 model
      (1) Scenario A1 Globalized Rapid Economic Growth
      (2) Scenario B1 Globalized Environmental Sustainability
   b. NC AR PCM model
      (1) Scenario A1 Globalized Rapid Economic Growth
      (2) Scenario B1 Globalized Environmental Sustainability
4. Erosion Analysis based on Soil and Topography using
   a. A statistical combination of 8 climate models
      (1) Scenario A1b Globalized Rapid Economic Growth.

5.3 Visualizing the climate data Used

It is appropriate to more closely examine the downscaled data used in the research. Figure 17 shows the Canadian precipitation data for the years 2000, 2050 and 2099, and Figure 18 shows a combined image for temperature and precipitation for 2099. It would be well to pay particular attention to the part of the map that indicates that the very wet area along the Ohio Valley is projected to disappear into the Atlantic Ocean off the eastern coast of Florida during the next century. Among other observations, this trend has a good deal of significance to the Army installations in this region.

5.4 Approach to manipulation of the climate change prediction data

5.4.1 Bailey's ecosystems change in response to climate data

The Bailey map was generated by experts who drew polygon shapes on a paper map. This means that although there was significant knowledge behind the work, it was also generalized. Realizing this, our task was to show climate change affecting those ecosystems. We chose to base our work on the well known version of Bailey's Ecoregions from the US Digital Atlas.* We also chose to deal with the units at their most detailed level, known as "Sections" (see Section 3.1.1 for further detail).

---

* http://www-atlas.usgs.gov/metadata/ecoregp075.html (file name: ecoregp075) [BAD LINK].

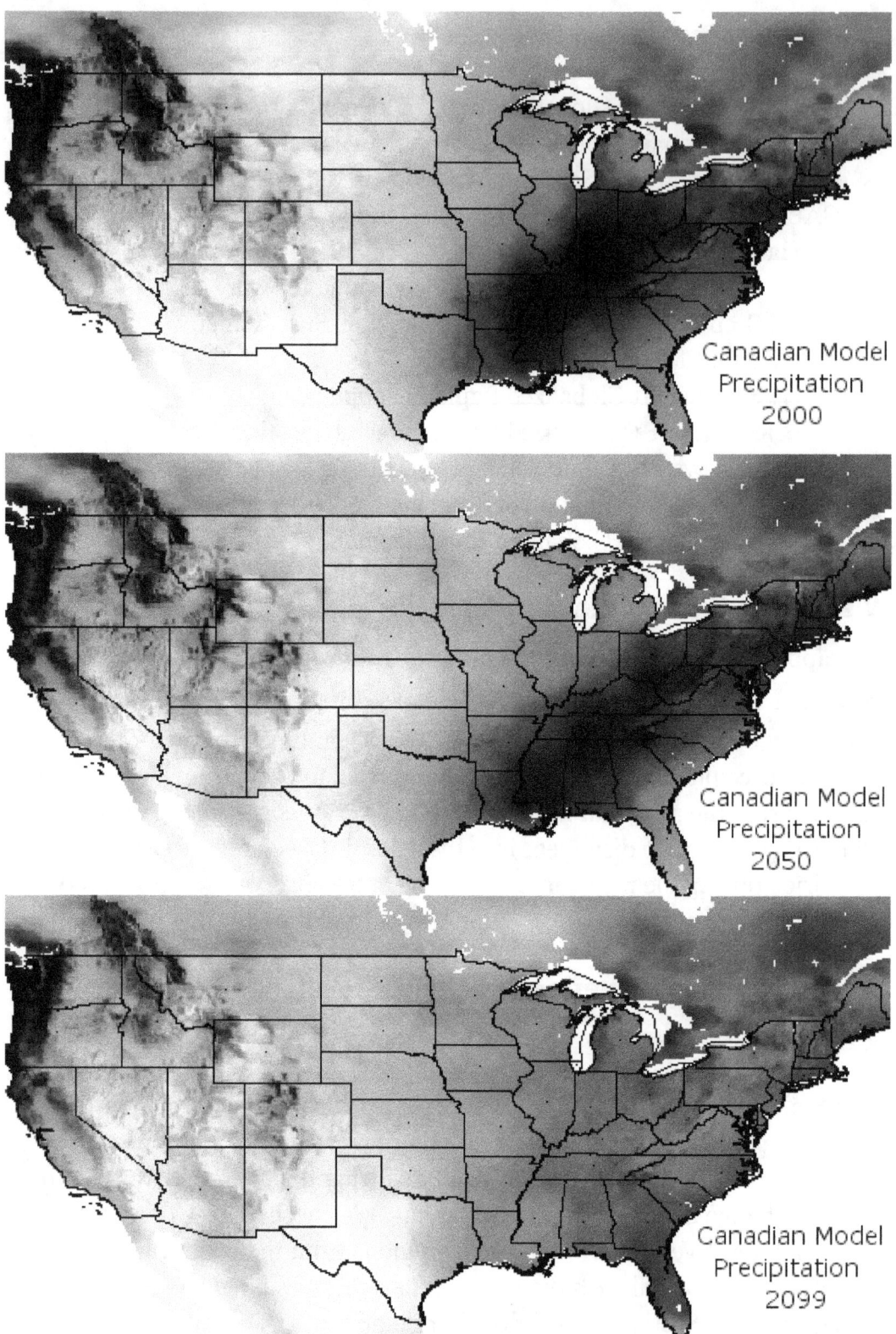

Figure 17. General view of raw Canadian Model Precipitation (Canadian CGCM 3.1.5
Scenario A1b). The very wet area along the Ohio Valley disappears, Western Texas becomes
drier, and Southern Arizona becomes slightly less arid.

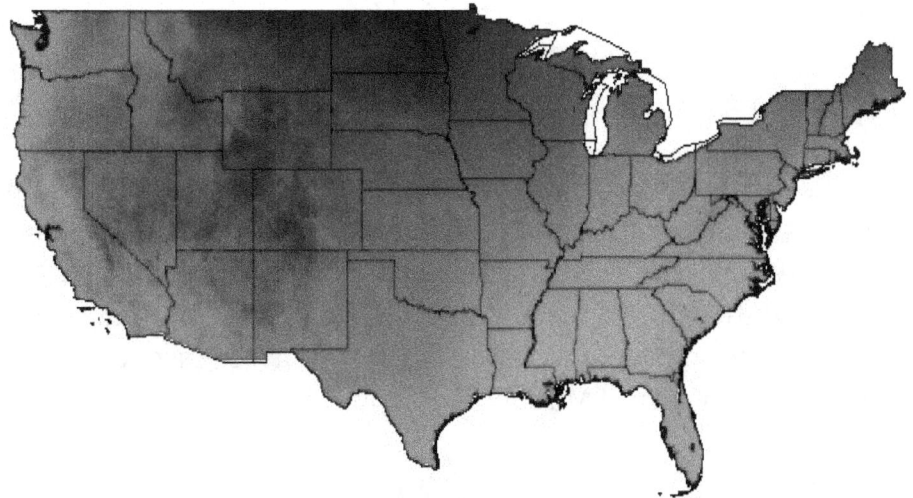

Figure 18. Combined image for temperature and precipitation for 2099 (CGCM 3.1.5Scenario A1b; similar for CGCM 3.1.5Scenario B1). Brighter red means higher temperature (so darker red is lower), brighter blue means higher precipitation (so darker is lower).

Although not the most recent, we used the standard version of Bailey's because that is the description with which most individuals are familiar. It fit with our thrust because the ecosystem sizes were more compatible with the sizes of the 78 different classifications we were able to generate using the combined precipitation and temperature unsupervised classification. Since 78 unsupervised classes became our limiting factor, trying to match that with a much more recent and detailed map would not have resulted in a net information gain. Realizing the limits of the 78 unsupervised classes for both Bailey's and Omernik's, we chose to adopt the more comprehensive similar approach with the Hargrove approach.

Since we wished to follow changes in the ecosystem over time, it was necessary to correlate the Baileys map (Figure 19) with the items that would change over time, namely temperature and precipitation, and specifically with the classified year 2000 temperature/precipitation map (Figure 20). First the vector Bailey's map was converted to a raster that coordinated in resolution with the downscaled data (Figure 21).

After conversion to raster, correlation statistics between the Resultant Classified Image (Figure 20) and the original Bailey's delineation (Figure 19) were generated. Finally, those categories most in common with both were used to reclassify the temperature and precipitation map into an equivalent ecosystem map (Figure 21) for the revised Bailey's map, which could then be used with the climate data to project changes over time (Figures 22 and 23).

The original Baileys map and the resultant classified image are certainly different, but the later does match the former roughly. In general one can see that the systems shift a considerable amount, but tend to move approximately southward. For example, while the classic *Columbia Basin* becomes a smattering of points in the northwest, a large section of similar climate develops in central Missouri by 2099.

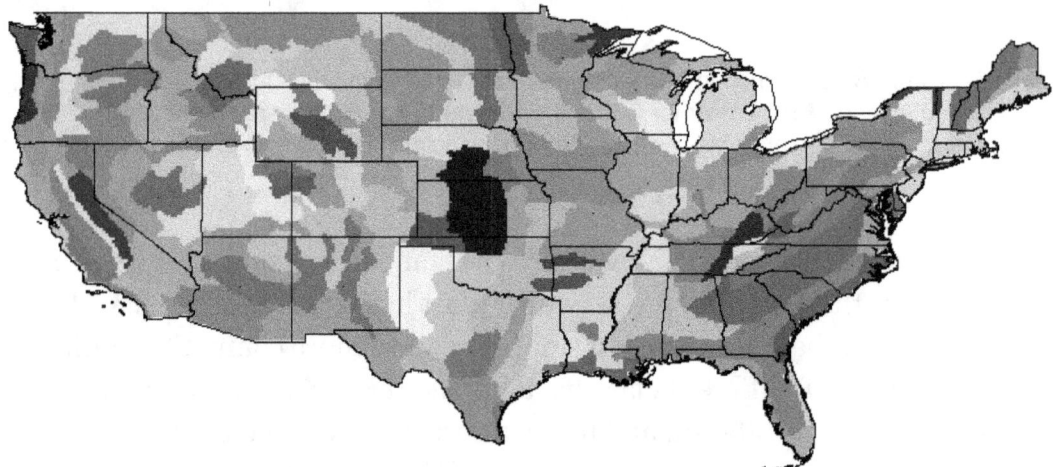

Figure 19. Baileys original regions.

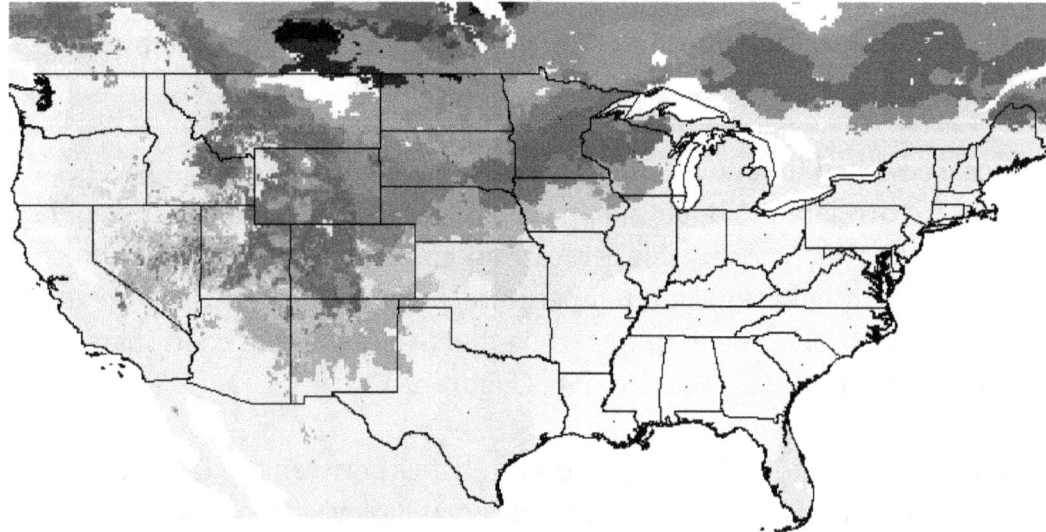

Figure 20. Resultant classified image (78 classes) for 2000 for precipitation and temperature (CM3_1.5 Scenario A1b).

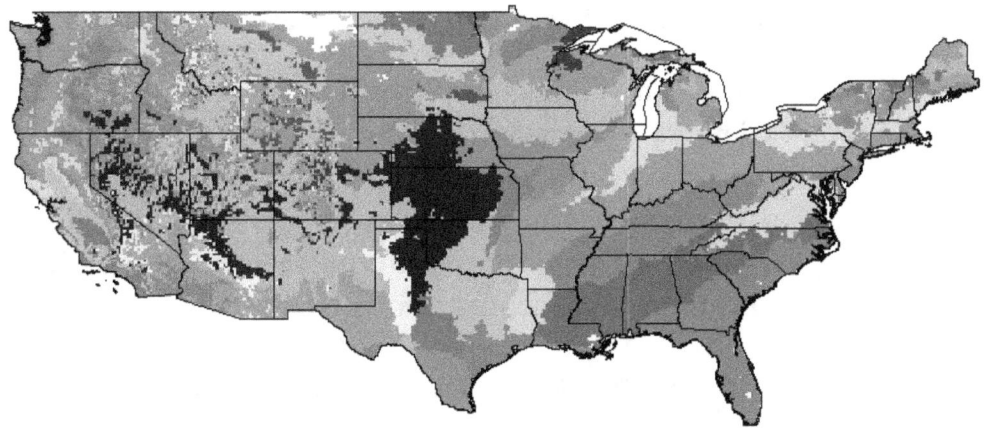

Figure 21. Revised Bailey's ecoregions based on 2000 temperature and precipitation and using the same color table as above (CM3_1.5 Scenario A1b).

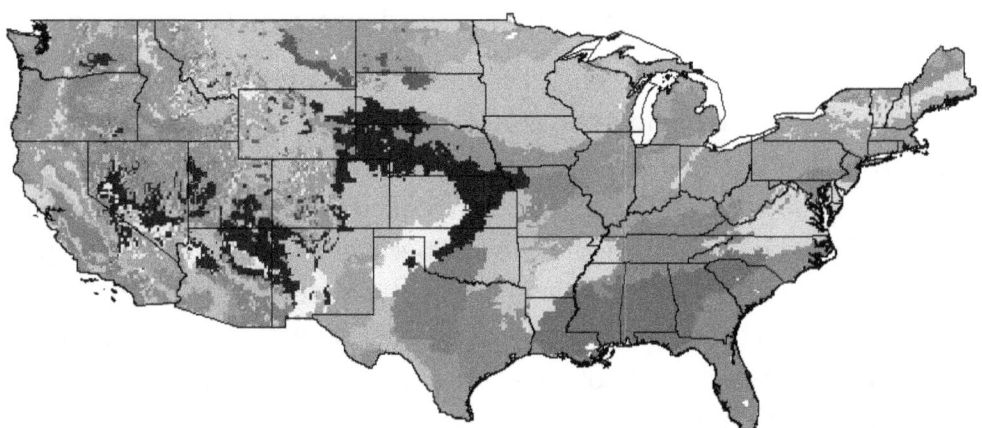

Figure 22. Bailey's ecoregions based on 2050 temperature and precipitation and using the same color table (CM3_1.5 Scenario A1b).

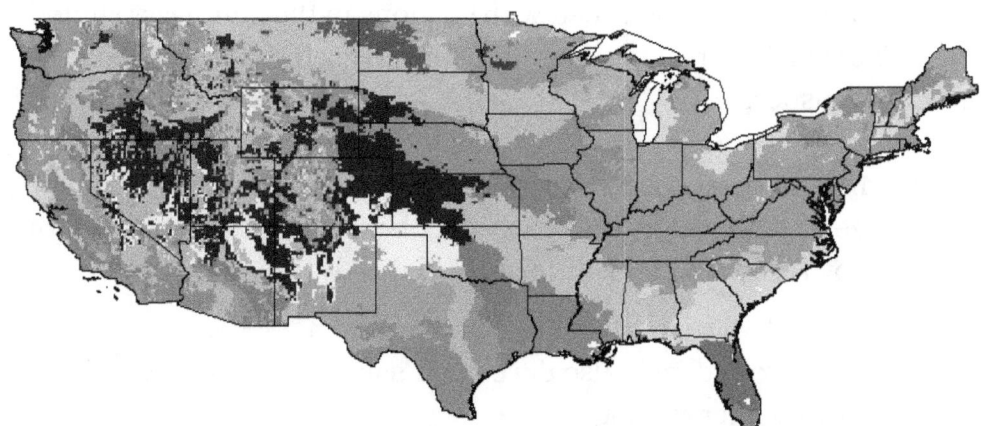

Figure 23. Bailey's ecoregions based on 2099 temperature and precipitation and using the same color table (CM3_1.5 Scenario B1).

5.4.2 Omernik's ecosystems change in response to climate data.

We used a similar procedure to roughly follow the changes one would expect in Omernik's definition of ecosystems. Our data source was from the "useco_3_areas" file obtained from the National Atlas.* In a broad sense, the resultant map from our Omernik-based analysis reflects the original well; however, Washington, Oregon and California seem poorly classified and changes based in those areas should be treated carefully.

5.4.3 Evaluation of reclassification of classical ecosystem maps.

Although our procedure is sound, the fact that we were only able to generate 78 imagery classes covering the whole United States is a severe limiting factor. Each of these 78 class regions covers an area close to the size of an average state. It is suggested that all the classifications would be improved by:

- Ensuring many more categories could be generated so that there are many units per ecoregion polygon. Consequently, the likelihood of aerial coverage matching well will increase.
- Ensuring the generated categories are based on more layers. For example, it should be possible from the temperature and precipitation files to generate additional layers such as frost free season, percent of year that is in dry season, percent of year of high temperatures, etc. These additional layers should not repeat the information in the temperature and precipitation files.

Since the temperature/precipitation patterns along the Pacific Northwest are similar to the Southeast, we can expect the Pacific Northwest to be confused with (or to be identified with) regions in the Southeast United States. Therefore, ecoregions identified for the Pacific Northwest should be considered suspect.

Since we are also using the Hargrove data in this research, it is useful to characterize its similarities and differences between our analysis and that conducted by Hargrove and Hoffman (2004):

- Similarities
  - The Hargrove data is also developed using the Unsupervised Imagery Classifier technique.

---

* http://www.nationalatlas.gov/mld/ecoomrp.html

- o We both used the A1 and B1 scenarios for our work.
- o We both normalized our imagery layers before the analysis was run so that each layer had equal weight.
- Differences
  - o Hargrove's work has more time and funding so it is much more detailed than ours.
  - o They standardized their resolution at 1km while our best climate data is a 1/8th degree (about 10km).
  - o They used 17 variables to make their 30,000 categories across the world while we used only 2 to make 78 categories over the CONUS area.
  - o They used the Hadley and PCM models for their predictions while we used the Canadian and CIRSO models.

There is no question that our predictive Bailey- and Omernik-based maps do not exactly match the original Bailey's and Omernik's maps. It is in part due to this concern issue we also adopted the Hargrove work. It should be noted that when completed, a comparison of our simple approach results with his more complicated approach results indicated that the major conclusions did not change. Comparing the consistency of results of the various approaches was a core concept in carrying out this study.

5.4.4 Ecosystem changes based on Hargrove data

In terms of the Hargrove data, we developed future habitat maps for the CONUS based on forecasts from GCMs and habitat classifications developed by the GAP program as correlated with the Hargrove maps (Figure 24) based on the Hadley Centre model (HadCM3) and the National Center for Atmospheric Research (NCAR) Parallel Climate Model (Westervelt and Hargrove 2010). The Hargrove approach applies the Multivariate Geographic Clustering (MGC) procedure simultaneously using 9 sets: one representing the current global state and eight representing forecasted future states. Each ecoregion map included 30,000 unique clusters representing eco-units based on 17 input map layers. With these clusters, we reclassified the data to generate our analyses and in more detail in a separate sister report (Westervelt and Hargrove 2010).

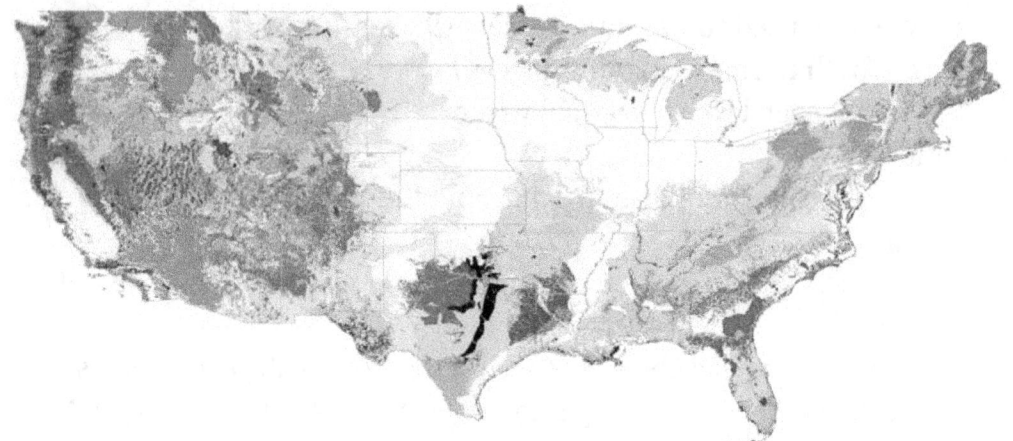

Figure 24. Hargrove's global ecosystem map reclassified to GAP categories
as used in this report.

5.5   Approach to climate change effects on erosion

5.5.1 The link between precipitation, event intensity, and erosion

As mentioned earlier, our primary GCM inputs are limited to temperature and precipitation changes between now and 2099. However, fluvial erosion increases as a rainfall event increases in intensity. So when considering erosion potential, the issue is not if overall precipitation increases, but if the intensity of precipitation (specifically distinct rainfall events) increases.

The USDA has long studied the processes of soil erosion, notably through its Agricultural Research Service (ARS). The ARS utilizes various predictive models to understand potential erosion, most notably the Revised Universal Soil Loss Equation (RUSLE) and the Water Erosion Prediction Project (WEPP) (Renard et al. 1997, p 14). The erosion rate for a given site results from the combination of multiple physical and management variables; these factors can be expressed in an equation of the form:

$$E = f(C, S, T, SS, M)$$

where:

> E = erosion
> F = function of ()
> C = climate
> S = soil properties
> T = topography
> SS = soil surface conditions
> M = human activities/management practices.

"RUSLE is an erosion model designed to predict the longtime average annual soil loss carried by runoff from specific field slopes in the specified cropping and management systems as well as from rangeland" (Renard et al. 1997, p 14). RUSLE and its predecessor (USLE) compute the average annual erosion expected as:

$$A = R * K * L * S * C * P$$

where:

A = computed spatial average soil loss
R = rainfall runoff erosivity factor
K = soil erodibility factor
L = slope length factor
S = slope steepness factor
C = cover-management factor
P = support practice factor.

When considering the potential effects of climate change on erosion, the R-factor within RUSLE is the most appropriate focus. While other characteristics may be altered, the R-factor will be the most significantly modified due to climate change. In fact, data has indicated that when factors other than rainfall are held constant, soil losses from cultivated fields are directly proportional to a rainstorm parameter: the total storm energy (E) times the maximum 30-minute intensity (I_{30}). "The EI is a statistical interaction term that reflects how total energy and peak intensity are combined in each particular storm .... The relation of soil loss to the EI parameter is assumed to be linear, and the parameter's individual storm values are directly additive" (Renard et al. 1997, p 23). While it was originally concluded that significant erosion was associated with only a few severe storms (that it is solely a function of peak intensities) more than 30 years of measurements in numerous states have shown that to accurately estimate average annual soil loss calculations must include the cumulative effects of the many moderate-sized storms as well as the effects of the occasional severe ones (Renard et al. 1997, p 22).

Fortunately, research dealing with the effects of climate change on erosion has become increasingly widespread in the 21st century. Many studies focus on a specific place, such as the Midwestern United States or the Loess Plateau of China, but some look at specific characteristics such as semi-arid areas or cropland. To this point, studies have focused on a place or region joined by a common characteristic (usually geological or climate); no studies have yet been published that provide a comprehensive understanding of multiple regions and the various reactions to the effects of climate change. In 2009, Wei et al. published a review of the "effects of rainfall change on water erosion processes in terrestrial ecosystems." The comprehensive review provides a survey of the most important science in regard to climate-induced precipitation change and its influence on soil erosion. Table 5 summarizes Wei's findings.

It is widely agreed that global climate change will lead to fluctuations in annual precipitation. In 2003, the Soil and Water Conservation Society (SWCS) published a report that reviewed existing literature and engaged a panel of experts to answer the question of how precipitation predictions due to climate change would affect soil erosion and runoff on cropland.

The report ultimate concluded that, "upward trends in total precipitation, coupled with a bias toward more extreme precipitation events, are indicated in both simulated and observed climate regimes," and that "regional, seasonal, and temporal variability in precipitation is large in both simulated and observed climate regimes" (SWCS 2003). Studies that look into the precipitation patterns from the 20th century consistently reaffirm the IPCC's claim of increasingly variable precipitation. Over the course of the 20th century, there was a definite increase in annual precipitation, and a significant portion of that increase came in the form of extreme rainfall events. Between 1910 and 1996, precipitation in the United States increased by around 10% and of that increase 53% came in the form of heavy and extreme daily rainfall events exceeding 2 inches (Karl and Knight 1998). The precipitation pattern of the 20th century is expected to continue (or even accelerate) as both the number of days with precipitation and the percentage of precipitation in the form of extreme (or heavy) rainfall events increase.

The IPCC reports that "some studies project widespread increases in extreme precipitation, with greater risks of not only flooding from intense precipitation, but also droughts from greater temporal variability in precipitation" (Field et al. 2007) According to the IPCC and others, projections generally show that changes in extreme precipitation will be much larger than changes in mean precipitation.

Table 5. Major precipitation intensity and erosion studies.

Major Findings and Research Regarding Rainfall Change and Water Erosion Dynamics Around the World

*Reproduced from Wei et al., Effects of rainfall change on water erosion processes in terrestrial ecosystems: a review. Progress in Phuysical Geography. 2009

Covered Geographical Area	Major Conclusions/Findings	Methodology	Source
Midwest USA	10-310% increase in runoff and 33-274% increase in erosion due to increased rainfall and reduced land coverage	Water Erosion Prediction Project (WEPP) model	O'Neal et al., 2005
Meuse basin, Europe	3% increase in rain erosivity inducing 333% increase in water erosion	WATEM/SEDEM model	Ward et al., 2009
South Korea	20% increase in storm depths and occurance causing 54-60% and 27-62% increase in runoff and soil loss, respectively	Climate generator (CLIGEN); Water Erosion Prediction Project (WEPP) model	Kim et al., 2009
Saxony, Germany	22-66% increase in erosion due to increased intensity and extreme events	ECHAM4-OPYC3 and EROSION2D model	Michael et al., 2005
Brazil	22-33% increase in mean annual sediment yield caused by 2% increase in annual rainfall	Hadley Center climate model (HadCM2)	Favis-Mortlock and Guerra, 1999
Different location in USA	Each 1% change in rainfall may cause 2% and 1.7% changes in runoff and erosion, respectively	CLIGEN model and regression equations	Pruski and Nearing, 2002
Global Scale	7% increase in rainfall during the twenty-first century	GCMs (general circulation models)	Houghton et al., 2001
South Downs, UK	7% increase in precipitation causing 26% increase in water erosion	Erosion Productivity Impact Calculator (EPIC) model	Favis-Mortlock and Boardman, 1995
Changqu tableland, Loess hilly area, China	23-37% increase in annual rainfall, 29-79% increase in runoff and 2-81% increase in soil erosion	HadCM3, WEPP and stochastic weather generator (CLIGEN)	Zhang and Liu, 2005
South Africa	A 10% increase in rainfall may lead to a 20-40% increase in runoff	CERES-Maize and ACRU models	Schulze, 2000
Loess Plateau, China	4-18% increase in rainfall with runoff increasing from 6% to 112% and erosion increasing from -10% to +167%	GCMs (general circulation models)	Zhang et al., 2008
Greece	The length and frequency of flood are predicted to increase twofold and threefold, respectively	Goddard Institute for Space Studies climate change model	Panagoulia and Dimou, 1997
Dingxi, Gansu province, northwestern China	Runoff and erosion rates under rainfall extremes were 2.68 and 53.15 times the mean ordinary rates, respectively	Statistics on long-term consecutive field data in situ	Wei et al., 2009
Global Scale	About 40% erosion potential due to increase precipitation	GIS-based RUSLE model	Yang et al., 2003

Along with land cover, soil characteristics, and topography, intensity of rainfall is strongly connected to the degree at which surface erosion occurs. In fact, "the annual water erosion and nutrient loss under extremely heavy rainfall exceeds 50% of the total in many environments" (Gao et al. 2005).This is substantiated by Albergel et al. (2004) who demonstrated that 50% of the erosion/sedimentation that occurred in the Kamech dam (in Tunisia) between 1994 and 2002 was caused by only three extreme precipitation events and the resulting floods (Raclot and Albergel 2006).

Many studies that explore the impact of climate change on erosion examine its influence on all or most of the major contributing factors (soil, landcover, topography and precipitation) and how a change in one will impact the rest. Some are concerned with the effects of climate change on agriculture and with the conservation measures that should be used to negate the increased erosion potential. The entire purpose of the 2003 SWCS report was to make scientifically backed suggestions to farmers interested in soil conservation. O'Neal, et al. (2005) explored the influence of climate change on the types of crops farmers are likely to plant in the US Midwest, and how that change in landcover will affect erosion in that region. Due to increased precipitation and decreasing cover from temperature-stressed maize, they predicted that erosion in the Midwest will increase between 10 and 310% depending on multiple simulation variables and a 2% precipitation increase.

Favis-Mortlock (1999) examined future erosion under intensive soy bean cultivation in the Mato Grosso area of Brazil finding that sediment yield could increase by 22–33% with a 2% increase in rainfall or decrease by 7-13% with a <2% drop in rainfall. Zhang published two articles in 2005 that attempted to model the potential impacts of crop production in the state of Oklahoma and the Loess Plateau of China. Each one predicted an increase in precipitation, the intensity of the rainfall events delivering that precipitation and an associated increase in soil erosion. Others, concerned with the interaction between fluctuating ground cover and fluctuations in precipitation in a non-agricultural setting, obtained similar results to those above.

Zhang et al. (2010) used temporal downscaling to estimate the Climate Generator (CLIGEN) input parameters to generate daily weather series representing future climates. They then used the CLIGEN outputs to calculate specific data such as the EI necessary for an accurate R-factor to include in a projected RUSLE calculation. Such a downscaling of GCM data has important implications for use in projected erosion due to climate change.

As the global climate changes during the course of the 21st century, the bulk of current research suggests that precipitation regimes will become increasingly extreme leading to longer periods of drought followed by more intense rainfall events. Vegetative landcover will suffer during the drought periods and will not provide sufficient erosion resistance during storms. Climate change results will be highly localized and no predictions should be universally applied. However, it is safe to summarize that the body of research predicts a positive correlation between precipitation intensity and potential erosion in the 21st century. The procedure used in our analysis of projected erosion potential is based on both the strongly established soil science of RUSLE and GCM-based precipitation intensity projections.

5.5.2 Erosion section bibliography

Albergel, J., S. Nasri, J. M. Lamache`re. 2004. HYDROMED – Programme de recherche sur les lacs collinaires dans les zones semi- arides du pourtour me ´diterrane ´en. Revue des Sciences de l'Eau 17 (2):133-151.

Carter, T. R., R. N. Jones, X. Lu, S. Bhadwal, C. Conde, L.O. Mearns, B.C. O'Neill, M. D. A. Rounsevell, and M. B. Zurek. 2007. New assessment methods and the characterisation of future conditions. Climate Change 2007: Impacts, Adaptation and Vulnerability. Contribution of Working Group II to the Fourth Assessment Report of the Intergovernmental Panel on Climate Change. Cambridge, United Kingdom (UK): Cambridge University Press. 133-171.

Christensen, J. H., B. Hewitson, A. Busuioc, A. Chen, X. Gao, I. Held, R. Jones, R.K. Kolli, W.-T. Kwon, R. Laprise, V. Magaña Rueda, L. Mearns, C.G. Menéndez, J. Räisänen, A. Rinke, A. Sarr, and P. Whetton. 2007. Regional climate projections. in Climate Change 2007: The Physical Science Basis. Contribution of Working Group I to the Fourth Assessment Report of the Intergovernmental Panel on Climate Change. Cambridge, UK and New York, NY, USA: Cambridge University Press.

Favis-Mortlock D, Boardman J. 1995. Nonlinear responses of soil erosion to climate change: A modelling study on the UK South Downs. *Catena*. 25(1-4):365-387.

Favis-Mortlock D., Guerra, A. 1999. The implications of general circulation model estimates of rainfall for future erosion: A case study from Brazil. *Catena*. 37(3-4):329-354.

Field J. P., D. D. Breshears, J. J. Whicker. 2009. Toward a more holistic perspective of soil erosion: Why aeolian research needs to explicitly consider fluvial processes and interactions. *Aeolian Research*. 1(1-2):9-17.

Field, C. B., L. D. Mortsch, M. Brklacich, D. L. Forbes, P. Kovacs, J. A. Patz, S. W. Running, and M. J. Scott. 2007. North America. Climate change 2007: Impacts, adaptation and vulnerability. Contribution of working group II to the fourth assessment report of the Intergovernmental Panel on Climate Change. Cambridge, UK: Cambridge University Press.

Gao, C., J. Zhu, Y. Hosen, J. Zhou, D. Wang, L. Wang, and Y. Dou. 2005. Effects of extreme rainfall on the export of nutrients from agricultural land. *Acta Geographica Sinica*. 60:991–97 (in Chinese with English abstract). Cited in: W. Wei, L. Chen, and B. Fu. 2009. Effects of rainfall change on water erosion processes in terrestrial ecosystems: a review. *Progress in Physical Geography*. 33(3):313.

Karl, T. R., and R. W. Knight. 1998. Secular trends of precipitation amount, frequency, and intensity in the United States. Bulletin of the American Meteorological Society. 79(2):231-241.

Meehl, G.A., T.F. Stocker, W.D. Collins, P. Friedlingstein, A.T. Gaye, J.M. Gregory, A. Kitoh, R. Knutti, J.M. Murphy, A. Noda, S.C.B. Raper, I.G. Watterson, A.J. Weaver and Z.-C. Zhao. 2007. Global Climate Projections. In: Climate Change 2007: The Physical Science Basis. Contribution of Working Group I to the Fourth Assessment Report of the Intergovernmental Panel on Climate Change [Solomon, S., D. Qin, M. Manning, Z. Chen, M. Marquis, K.B. Averyt, M. Tignor and H.L. Miller (eds.)]. Cambridge University Press, Cambridge, United Kingdom and New York, NY, USA.

Nearing, M. A., V. Jetten, C. Baffaut, O. Cerdan, A. Couturier, M. Hernandez, Y. Le Bissonnais, M. H. Nichols, J. P. Nunes, C. S. Renschler, V. Souchere, and K. Van Oost. 2005. Modeling response of soil erosion and runoff to changes in precipitation and cover. *Catena*.61(2-3):131-154.

Nunes J. P., J. Seixas, J. J. Keizer, and A. J. Ferreira. 2009. Sensitivity of runoff and soil erosion to climate change in two Mediterranean watersheds. Part II: Assessing impacts from changes in storm rainfall, soil moisture and vegetation cover. *Hydrological Processes*. 1220(March):1212- 1220.

Nunes P., N.R. Pacheco. 2008. Vulnerability of water resources, vegetation productivity and soil erosion to climate change in Mediterranean watersheds. Hydrological Processes. 3134(October 2007):3115- 3134.

O'Neal M, Nearing M, Vining R, Southworth J, Pfeifer R. Climate change impacts on soil erosion in Midwest United States with changes in crop management. *Catena*. 2005;61(2-3):165-184.

Pruski F., and M. Nearing. 2002. Climate-induced changes in erosion during the 21st century for eight US locations. Water Resources Research. 38(12).

Raclot, D., and J. Albergel. 2006. Runoff and water erosion modelling using WEPP on a Mediterranean cultivated catchment. Physics and Chemistry of the Earth, Parts A/B/C. 31(17):1039.

Renard, K. G., G. R. Foster, G. A. Weesies, D. K. McCool, and D. C. Yoder (coordinators). 1997. Predicting Soil Erosion by Water: A Guide to Conservation Planning With the Revised Universal Soil Loss Equation (RUSLE). Agricultural Handbook No. 703. Washington, DC: US Department of Agriculture (USDA), p 14.

Soil and Water Conservation Society (SWCS). 2003. Conservation Implications of Climate Change: Soil Erosion and Runoff from Cropland. Ankeny, IA: SWCS.

Wei W, Chen L, and L. Y. Fu B. 2009. Responses of water erosion to rainfall extremes and vegetation types in a loess semiarid hilly area, NW China. Hydrological Processes. 1791(May):1780- 1791.

———. 2009. Effects of rainfall change on water erosion processes in terrestrial ecosystems: A review. Progress in Physical Geography. 33(3):307-318.

Yang D, S. Kanae, T. Oki, T. Koike, and K. Musiake. 2003. Global potential soil erosion with reference to land use and climate changes. Hydrological Processes. 17(14):2913-2928.

Zhang X., and W. Liu. 2005. Simulating potential response of hydrology, soil erosion, and crop productivity to climate change in Changwu tableland region on the Loess Plateau of China. *Agricultural and Forest Meteorology*. 131(3-4):127-142.

Zhang X., Liu W., Z. Li, and F. Zheng. 2009. Simulating site-specific impacts of climate change on soil erosion and surface hydrology in southern Loess Plateau of China. *Catena*. 79(3):237-242.

Zhang X., and M. Nearing. 2005. Impact of climate change on soil erosion, runoff, and wheat productivity in central Oklahoma. *Catena*. 61(2-3):185-195.

Zhang X. 2007. A comparison of explicit and implicit spatial downscaling of GCM output for soil erosion and crop production assessments. Climatic Change. 84(3-4):337-363.

Zhang Y-G., M.A. Nearing, X.-C. Zhang, Y. Xie, and H. Wei, 2010. Projected rainfall erosivity changes under climate change from multimodel and multiscenario projections in Northeast China. Journal of Hydrology. 384:97-106.

Zhou, Z., Z. Shangguan, and D. Zhao. 2006.Modeling vegetation coverage and soil erosion in the Loess Plateau Area of China. *Ecological Modelling*. 198(1-2):263-268.

5.5.3 The process for erosion risk analysis

Our erosion risk analysis is an attempt to use projected precipitation intensity data to determine possible changes in erosion rates using RUSLE. There is direct relationship between the projected precipitation intensity data used in this analysis and RUSLE's R-factor (i.e., an increase in precipitation intensity results in an increase in R-factor). However, a 5% increase in our precipitation intensity data will not represent a 5% increase in R-factor. The variation arises due to the fact that R-factor is calculated from detailed information based on observations of past events while the data used in this analysis is based on low resolution projections of future events.

There are methods available to project the detailed storm data needed to estimate a future R-factor that will be explained later in this section. However, the goal of this analysis is to provide a rank order of future erosion potential due to projected changes in precipitation intensity. Since this analysis does not use RUSLE to estimate specific levels of erosion, using projected change in precipitation intensity as an indication of changes in R is acceptable.

For the purpose of this analysis, we assume that all other factors (K, L, S, C, and P) will remain constant throughout the analysis period. This allows us to isolate the impacts of climate change as separate from other shifts in landcover or management that might affect erosion potential. This relationship can be expressed in an equation of the form:

$$A_{Current} = R_{Current}(KLSCP)$$
$$A_{Future} = R_{Future}(KLSCP)$$

where:

A	=	computed spatial average soil loss
R	=	rainfall runoff erosivity factor (in this case, projected change in precipitation intensity)
K	=	soil erodibility factor
L	=	slope length factor
S	=	slope steepness factor
C	=	cover-management factor
P	=	support practice factor.

The precipitation intensity data used in this study were generously shared by Dr. Claudia Tebaldi, a research scientist for Climate Central, Inc. The multi-model ensemble (eight models total) used to create the dataset comprises PCM, CCSM3, GFDL-CM2.0, GFDL-CM2.1, MIROC3.2-hires, MIROC3.2-medres, CNRM-CM3, and INMCM3_0. The values in the data represent the Simple Daily Intensity Index (SDII), which is defined as the total annual precipitation amount divided by the total number of wet days in the year. Tebaldi et al. have shown that the SDII "is qualitatively representative of other precipitation extremes indices, and all the change in precipitation extremes indices conform to roughly the same pattern. Therefore, the SDII is "useful as a starting point to analyze processes associated with producing the geographic patterns of change in precipitation extremes" (Meehl et al. 2005, 1). The data are presented as a mean "daily" precipitation. Each year is assigned only one value, but the value represents an average wet day in a given year and therefore is a "daily" not an "annual" measure. Tebaldi et al. calculated the mean SDII for both 1980–1999 and 2080–2099. They then calculated the difference between the means for the two 20-year periods and thus calculated a dataset that represents the projected change between the average precipitation intensity at the end of the 20th century (1980–1999) and the average projected precipitation intensity at the end of the 21st century (2080–2099).*

For the purposes of this analysis we had the option of using one of two sets of values based on the same data. The first was the absolute change in precipitation intensity (expressed in mm/day). The second set was based on the absolute change, but had been rescaled in terms of the climatological standard deviation. Because the standard deviation was calculated based on values across the entire planet, we have elected to use the absolute change values. The SDII data obtained from Dr. Tebaldi is on the large scale often associated with GCMs (~2.8 degree); therefore, we sampled the data down to a smaller grid (1.0 decimal degrees) using a cubic resampling technique to obtain a greater range of values across the installations (Figure 25).

---

* The dataset is nearly identical to the one used in Meehl, Arblaster, and Tebaldi (2005). The primary difference is this dataset includes eight models whereas the data used in the article included nine.

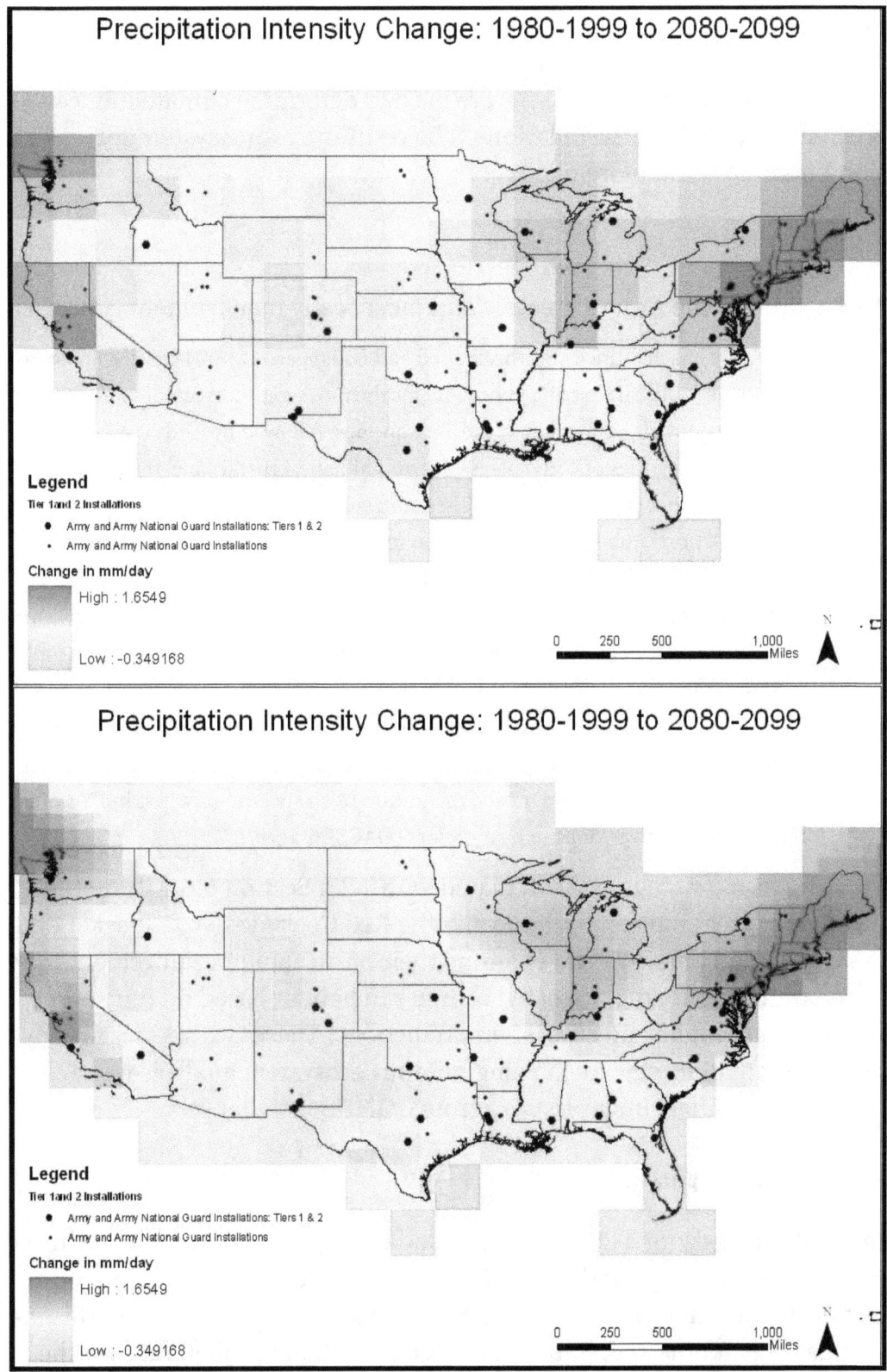

Figure 25. The first image shows the original data with a 2.8-degree resolution. The second
image shows the same data after resampling to a 1-degree resolution.

Final statistics on each individual installation were teased from the precipitation intensity map using a zonal statistics tool. The results from the procedure (outlined in section 6.5) were used to determine climate induced erosion levels at Army installations. The resulting rating system for erosion potential is therefore relative.

5.5.4 TES

TES are one of the Army's greatest and most costly management concerns:

> The Army has identified 188 threatened and endangered (T&E) species on 99 installations for fiscal year (FY) 2007. By far the most common category is plants, which account for 62 percent of the T&E species, followed by birds (14 percent). The other categories of T&E species are amphibians, crustaceans, fish, insects, mammals, other invertebrates, reptiles, and snails.
>
> During FY 2007, the Army spent $45.09 million on T&E species management. This sum included Endangered Species Act (ESA) reportable expenditures of $35.83 million and non-reportable expenditures of $9.26 million. These amounts do not include expenditures by military operations or military construction for work-arounds and avoidance.
>
> The 10 installations with the highest reportable expenses accounted for 68 percent of the Army's total reportable expenses. The red-cockaded woodpecker and desert tortoise required the most expenditure of all species—not just in FY 2007, but also cumulatively for the past 5 years (OACSIM 2009).

To bring focus to the issues associated with TES, Section 6.6 will spotlight two TES of great interest to the Army: the Red Cockaded Woodpecker and the Gopher Tortoise. We will examine their basic habitat requirements and compare those with the projected changes in habitat based on our source climate change data and ecosystem predictions. These two species are intended as an initial step at showing how our ecosystem analysis work can be applied to other similar issues across CONUS installations.

5.5.5 Invasive species

In addition needs for TES on Army installations, the management of invasive species costs a great deal of time and money. Here we examine the effect of climate change on invasive species. Our primary resource in this investigation, and the best standard source of spatially explicit data on the subject, was an Army report, *US Army Installation Floristic Inventory Database* (HQUSACE 2007). From this report, installations were extracted and the number of invasive species/installation was summarized (Table 6). Unfortunately, invasive species populations are listed at only 18 installations, which represent only a small portion of the installations included in this study. It is assumed that these are the only installations for which surveys have been done.

From this limited data, we can say that, where surveys have been done, invasive species were found to be common, often ranging into the hundreds of species per installation. This information is only marginally useful for the purposes of predicting how the situation will evolve under the pressures of climate change over the next century. However, the fact that invasive species are common becomes the base understanding for dealing with the issue in terms of climate change.

Table 6. Summary of known invasive species.

Installation	Number of Invasive Species
Fort Benning	177
Fort Bliss	178
Fort Bragg	217
Fort Campbell	127
Fort Carson	54
Fort Drum	185
Fort Hood	196
Fort Hunter Liggett	135
National Training Center and Fort Irwin	45
Fort Knox	114
Fort Leonard Wood	182
Fort Lewis	195
Fort Polk	72
Fort Riley	182
Fort Rucker	130
Fort Sill	167
Fort Stewart	184
Fort Wainwright	52

6 Analyses and Results of Climate Change Effects on Ecosystems, Erosion, TES and Invasive Species at Army Installations

6.1 Objective

In this chapter our objective is to investigate what the changes will be to individual Army installations (roughly 128 were evaluated) and to highlight those locations expected to experience the greatest changes. The major sections will deal with precipitation amount and temperature change predicted by different models and different scenarios along with some indication as to what ecosystem changes and erosion impacts are to be expected. We limit the reporting to just two time horizons, 2000 and 2099 although the data is available for all years from 1950 to 2099. We also limit the installations named here (normally) to the top several in each evaluation because we wished to examine the stability of the results. Of course, nearby installations are likely to be similar in ranking even if not mentioned explicitly.

6.2 Precipitation

The following section simply presents the exiting downscaled climate change projection data and how that will relate to Army installations. No manipulation of the original downsized data has been carried out.

6.2.1 Canadian Model, Scenario A1b "Globalized Rapid economic Growth"

Table 7 lists the 10 Army installations of the 128 in this specific evaluation that the data indicated show the greatest precipitation change between the 2000 and 2099 data. (Appendix A lists the data from which this and all other evaluations in Sections 6.2.1 to 6.2.3 are extracted.) All of the changes indicate a decrease in expected precipitation. Although they range across the country, there is a clustering of many of these installations in the east-central United States near the Mississippi and Ohio River Valleys (Figure 26).

Table 7. Greatest precipitation change according to the Canadian Scenario A1b

Installation	Canadian Scenario A1b Prcp 20-99 Change	
	in./day	mm/day
Fort Knox	-0.319	-8.1
Fort Campbell	-0.307	-7.8
Milan Arsenal and Wildlife Management Area	-0.252	-6.4
Camp Atterbury Military Reservation	-0.248	-6.3
Mount Baker Helicopter Training Area	-0.248	-6.3
Redstone Arsenal	-0.220	-5.6
Pine Bluff Arsenal	-0.197	-5.0
Snoqualmie National Forest	-0.189	-4.8
Camp Joseph T. Robinson	-0.161	-4.1
Picatinny Arsenal	-0.150	-3.8

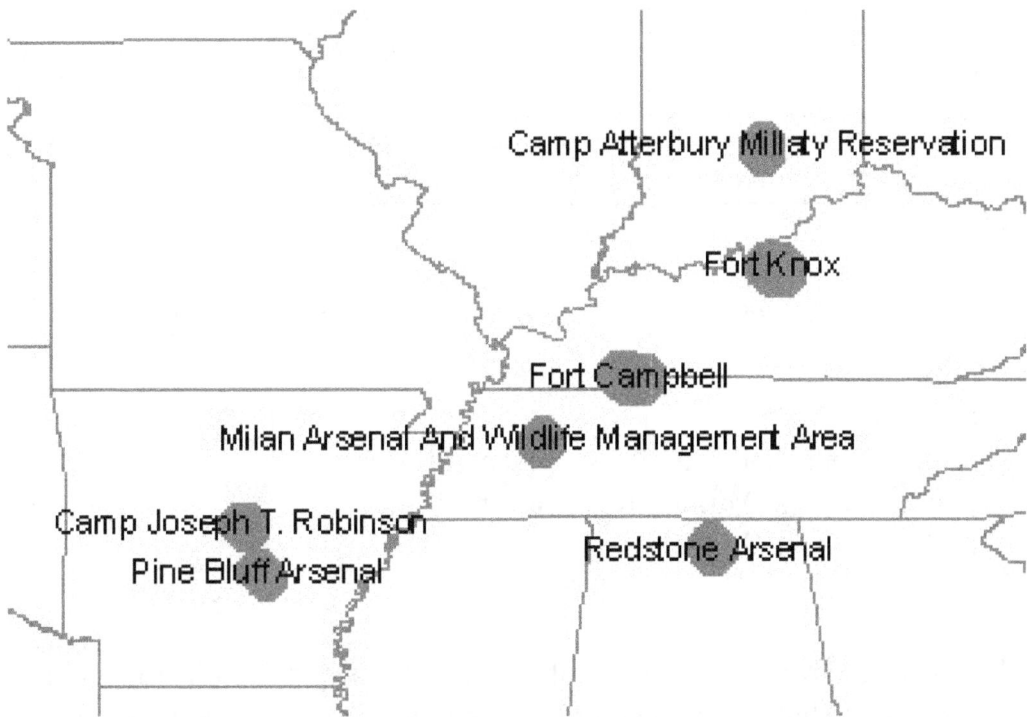

Figure 26. For decreased precipitation there is a clustering of Army installations based on the Canadian A1b scenario.

6.2.2 Canadian Model, Scenario B1 "Globalized Environmental Sustainability"

Table 8 lists those installations that experience the greatest change.

Table 8. Greatest precipitation change according to the Canadian Scenario B1.

Installation	Canadian Scenario B1 Prcp 20-99 Change in mm/day	
	in./day	mm/day
Fort Campbell	−0.354	−9.0
Milan Arsenal and Wildlife Management Area	−0.335	−8.5
Fort Knox	−0.311	−7.9
Hunter-Liggett Military Reservation	0.303	7.7
Redstone Arsenal	−0.276	−7.0
Hunter-Liggett Military Reservation	0.256	6.5
Pine Bluff Arsenal	−0.256	−6.5
Camp Atterbury Military Reservation	−0.232	−5.9
Presidio of Monterey	0.224	5.7
Camp Roberts Military Reservation	0.217	5.5

As with Canadian model scenario A1b, among the top installations *again exhibiting the greatest predicted decrease in precipitation* are:

- Fort Campbell
- Fort Knox
- Milan Arsenal and Wildlife Management Area
- Camp Atterbury Military Reservation
- Redstone Arsenal
- Pine Bluff Arsenal

This list includes six of the installations in Table 7 (p 59), indicating a high agreement of what will happen at these locations no matter what scenario is adopted. They are simply the most likely to be highly impacted.

6.2.3 Australian Model, Scenario A1b "Globalized Rapid Economic Growth"

Table 9 lists those installations that experience the greatest change in this scenario.

Table 9. Australian model, Scenario A1b

Installation	Australian Scenario A1b Prcp 20-99 Change	
	in./day	mm/day
Hunter-Liggett Military Reservation	0.417	10.6
Camp Parks Military Reservation	0.335	8.5
Presidio of Monterey	0.311	7.9
Los Alamitos Armed Forces Reserve Center	0.299	7.6
Camp Roberts Military Reservation	0.295	7.5
Fort Polk Military Reservation	−0.291	−7.4
Fort MacArthur	0.260	6.6
Redstone Arsenal	−0.220	−5.6
Sharpe General Depot (Field Annex)	0.185	4.7
Anniston Army Depot	−0.181	−4.6

It is interesting that those installations in the Canadian model, scenario B1 "Globalized Environmental Sustainability" (Table 8), predicted to have the greatest change in precipitation, also appear in the Australian model, scenario A1b (Table 9), namely:

- Hunter-Liggett Military Reservation – major increase
- Presidio of Monterey – major increase
- Camp Roberts Military Reservation – major increase
- Redstone Arsenal – major decrease.

Once again, this list of the top 8% in precipitation change among 128 installations indicate great stability in the precipitation predictions, particularly for Redstone Arsenal, which shows up in all three models.

As a matter of interest, we checked the rankings of the six **greatest predicted changes in precipitation** from Tables 7 and 8. In this section they ranked as:

- Fort Campbell – 47 of 128, or 37%
- Fort Knox – 82 of 128, or 64%
- Milan Arsenal and Wildlife Management Area – 39 of 128, or 30%
- Camp Atterbury Military Reservation – 70 of 128, or 55%
- Redstone Arsenal – 8 of 128, or 6%
- Pine Bluff Arsenal – 34 of 128, or 26%.

Of these, only Camp Atterbury and Fort Knox are not in the top 50% of the entire list that included Tables 7 and 8. The map in Figure 27 shows that these are the two most northerly of the highly impacted installations. The Australian model tends to place the bigger changes further to the south as all the other installations in the above list still belong in the top 50% change category.

Those installations showing the greatest decrease in predicted precipitation were mapped with the corresponding data from the Canadian and Australian models in Figure 27. Both models agree that this region will become drier and that that endpoint is roughly equivalent. The difference is due to the initial condition in the Canadian model suggests that the wet area is a little north of that shown in the Australian model. One would think that the 2000 data should be the same in both. However, the models usually start their predictions not at 2000, but usually at 1950 (historic data was not used in this report). Thus the year 2000 data is actually a prediction in climate models, not a starting point based on climate records observation. The purpose is to use the 50 years before 2000 as a test of the model's viability. If it does not pass that test reasonably it does not become a well recognized predictive model.

Since there appeared a consistency in those locations showing the greatest increase in predicted precipitation in Tables 7, 8, and 9, we mapped those installations in Figure 28. All of them are along the mid to southern California coastline.

6.3 Temperature

6.3.1 Canadian Model, Scenario A1b "Globalized Rapid Economic Growth"

Table 10 lists those installations that experience the greatest change in the globalized rapid economic growth scenario for the Canadian Model.

All of the changes indicate a large decrease in expected temperature. This is why the research area is termed climate change; the term "global warming" can be misleading. The group listed in Table 10 does not range across the country; rather, there is a clustering of installations in the south eastern United States that includes some of the Army's most important training and readiness installations (Figure 29). Further, these installations are concentrated in only three states: Alabama, Georgia, and North Carolina.

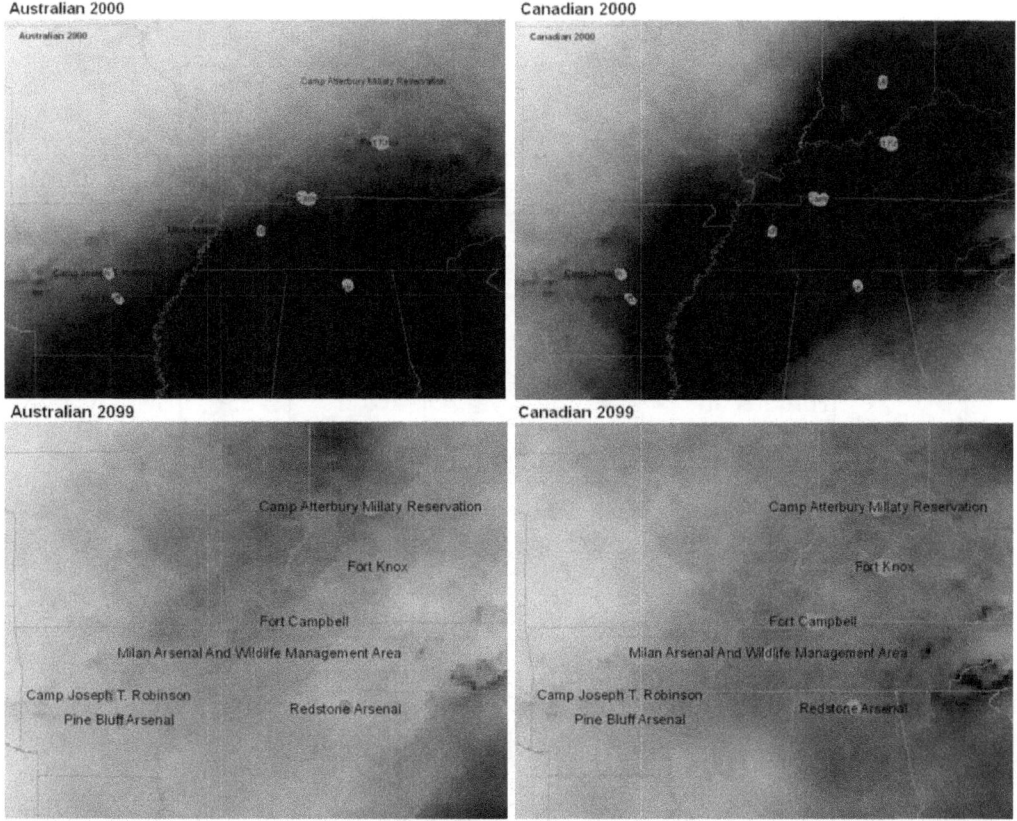

Figure 27. Comparison on precipitation models in the central Mississippi-Ohio River Valleys for 2000 and 2099. Darker means more rain. The Canadian model (right) predicts more rainfall in 2099 than the Australian model (bottom left), although the amount decreasing between 2000 and 2099 in the Canadian Model is greater.

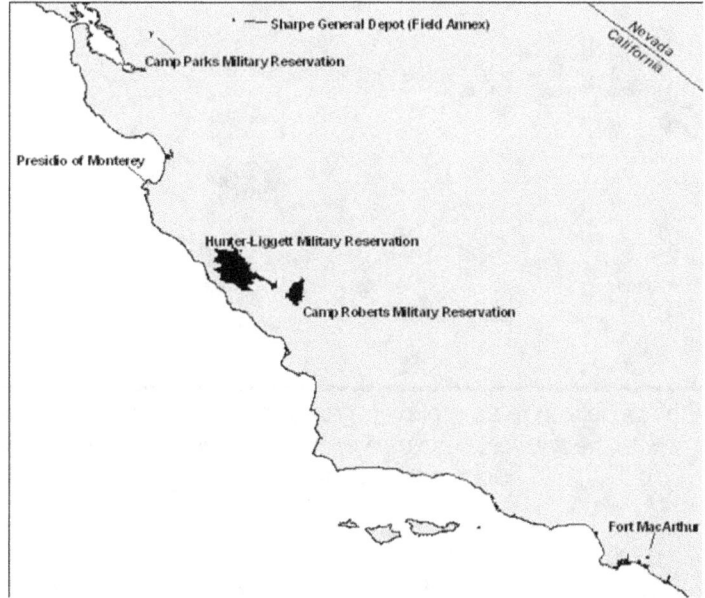

Figure 28. Installations consistently showing the greatest increased precipitation.

Table 10. Canadian Model (Scenario A1b), installations
that experience the greatest change in temperature.

Installation	Canadian Scenario A1b Temp 20-99 Change	
	°F	°C
Fort Rucker Military Reservation	−17.28	−9.6
Fort Benning Military Reservation	−16.74	−9.3
Fort Bragg Military Reservation	−16.02	−8.9
Fort Gillem Heliport	−16.02	−8.9
Fort Stewart	−16.02	−8.9
Fort McPherson	−15.84	−8.8
Camp MacKall Military Reservation	−15.84	−8.8
Hunter Army Airfield	−15.84	−8.8
Fort Gordon	−15.84	−8.8
Anniston Army Depot	−15.48	−8.6

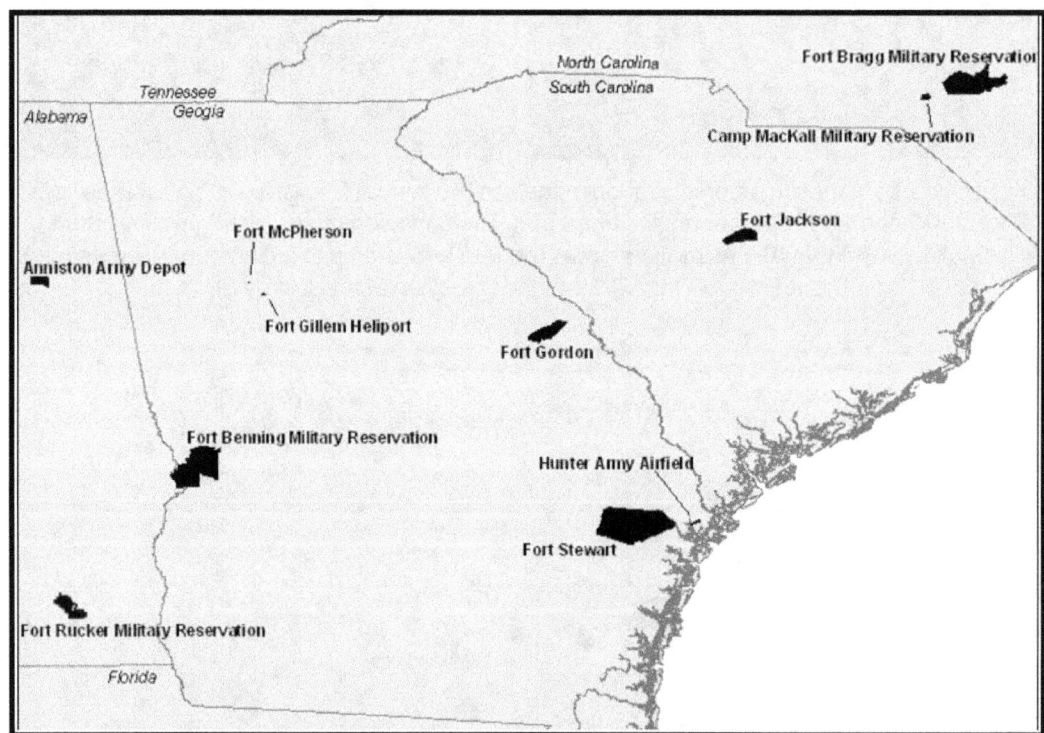

Figure 29. In the globalized rapid economic growth scenario for the Canadian Model, the
installations with the greatest change, a decrease in temperature all cluster in the
Southeastern United States.

It is interesting to note that the greatest precipitation changes occurred just to the west of this area where the temperatures are most decreased. An inspection of Figure 17 shows that in the coordinating precipitation model, the rainfall has not changed as much as the temperature. This indicates that perhaps the two concerns (temperature and precipitation change) are not directly connected.

6.3.2 Canadian Model, Scenario B1 "Globalized Environmental Sustainability"

The Army installations that experience the greatest change under a sustainability scenario are altogether different locations than the economic growth scenario. The increased temperatures seen here are almost in reverse degree to the decreased temperatures seen in the Canadian A1 model (Table 10). The installations listed in Table 11 reside first in the Montana/Utah area and second (with the exception of Buckley Air National Guard AF Base) in or near Texas (Figure 30). Though Texas and Oklahoma are not the worst impacted, their resident installations tend to be large and important military bases. In general, plains areas will observe higher temperatures. Military personnel will need more protection and vehicles will experience greater stress due to heat. In the B1 Scenario, it is valuable to understand what happened to the cooler Southeastern installations. Table 12 lists the B1 scenario data for same installations listed in Section 6.3.1 .

Table 11. Army Installations that experience the greatest change under Canadian Model, Scenario B1.

Installation	Canadian Scenario B1 Temp 20-99 Change	
	°F	°C
Fort William H. Harrison Military Reservation	16.74	9.3
Fort Wolters	16.20	9.0
Fort Sill Military Reservation	16.20	9.0
Bearmouth National Guard Training Area	14.94	8.3
Fort Hood	14.94	8.3
Camp Swift NG Facility	13.86	7.7
Buckley Air National Guard AF Base	13.68	7.6
Tooele Army Depot	13.14	7.3
Camp Bullis	13.14	7.3
Camp Williams	13.14	7.3

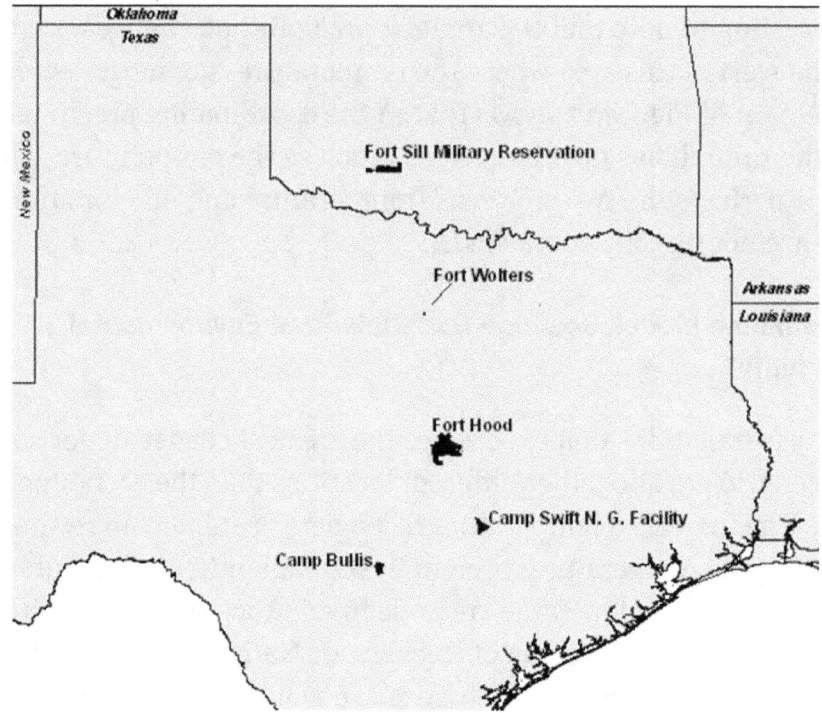

Figure 30. Warmer Texas area Installations.

Table 12. Canadian Model, Scenario B1, cooler installations.

Installation	Canadian Scenario B1Temp 20-99 Change	
	°F	°C
Fort Rucker Military Reservation	-10.26	-5.7
Fort Benning Military Reservation	-9.18	-5.1
Fort Bragg Military Reservation	-10.98	-6.1
Fort Gillem Heliport	-8.46	-4.7
Fort Stewart	-9.54	-5.3
Fort McPherson	-8.28	-4.6
Camp MacKall Military Reservation	-10.80	-6.0
Hunter Army Airfield	-9.36	-5.2
Fort Gordon	-8.82	-4.9
Anniston Army Depot	-7.92	-4.4

The southeastern installations still indicate temperature decreases, but at lower levels due to the fact that this is a sustainability scenario. In fact, Fort Bragg shows the greatest temperature decrease of any installation in the B1 Scenario. The salient point is that under the sustainability scenario, the most dramatic overall impacts to temperatures will be in those areas experiencing temperature increase rather than those areas in which the temperature will decrease.

Table 13. Installations with the greatest temperature change under Australian A1b scenario.

Installation	Australian Scenario A1b Temp 20-99 Change	
	°F	°C
White Sands Missile Range	11.52	6.4
Fort Bliss	11.16	6.2
Fort Bliss McGregor Range	10.98	6.1
Fort Wolters	10.80	6.0
Fort Hood	10.44	5.8
Camp Bullis	10.26	5.7
Fort Carson Military Reservation	10.08	5.6
Camp Swift N. G. Facility	10.08	5.6
Buckley Air National Guard AF Base	9.72	5.4
Natick Laboratories Military Reservation	9.72	5.4

6.3.3 Australian Model, Scenario A1b "Globalized Rapid Economic Growth"

Table 13 lists the installations that experience the greatest change under the Australian Rapid Growth Scenario. These locations show a temperature increase. The increase is not as great as the predicted by the Canadian model in the A1b scenario (Table 11). The list in Table 13 largely overlaps the Canadian Model B1 list (Table 12) and includes installations slightly to the west of those shown in Figure 30. The Canadian B1 and the Australian A1 scenarios indicate that the greatest impacts will occur in the western United States. In addition, the greatest change includes the same Western region as shown in Figure 30, with the addition of a group of installations to the north. Natick Laboratories is the only installation not in this general region and it is one of the least impacted listed in Table 13.

When we compare the Australian A1b temperature data directly to the comparable Canadian A1b scenario, what do the results of this model say happened to those installations the Canadian model claimed would be highly affected? Table 14 lists the installations originally reported from the Canadian A1b model in Table 10, but with the temperature change predictions for those installations derived from the Australian A1b model (Figure 31). The Australian model suggests that that Southeastern United States will experience little temperature change. The fact that these two models using the same scenario predict much different results suggests that more work needs to be done to make the predictions more consistent and reliable; obviously one cannot be correct. It was not within the scope of this research to address this further, but it is important to highlight.

Table 14. Comparison of Canadian and Australian A1b scenarios.

Installation	Australian Scenario A1b Temp 20–99 Change	
	°F	°C
Fort Rucker Military Reservation	0.36	0.2
Fort Benning Military Reservation	1.62	0.9
Fort Bragg Military Reservation	2.52	1.4
Fort Gillem Heliport	2.16	1.2
Fort Stewart	-0.18	-0.1
Fort McPherson	2.16	1.2
Camp MacKall Military Reservation	2.52	1.4
Hunter Army Airfield	0.18	0.1
Fort Gordon	1.80	1.0
Anniston Army Depot	2.70	1.5

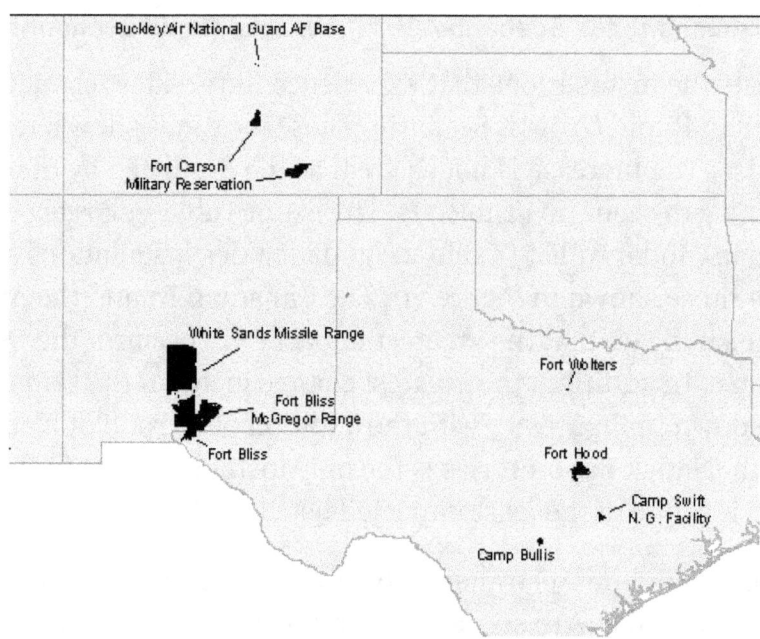

Figure 31. Greatest temperature change installations from the Australian model, scenario A1b.

An observation relating to all the discussions presented so far in Chapter 6 is appropriate. As we have pointed out, highly impacted installations often occur in clusters. Once you see this, it might be obvious that this should be the case because they lie within the regions that are highly impacted, so one might expect the observed groupings. The fact that this survey finds that to be true supports the both the logic is correct and that the predictions are largely consistent with a regional viewpoint.

6.4   Ecosystem changes

The results in this section are based on the analysis procedures outlined in Sections 5.4.1 to 5.4.4. It is very important to interpret these maps with the following caveats.

1. The Hadley and PCM models were chosen to represent relative extremes in GCM forecasts. The Canadian and Australian models were chosen to represent moderate forecasts. Similarly, the A1 and B1 gas-emission scenarios provide relative extremes in greenhouse gas emission rates over the 21st century.
2. Compared with the size of installations, the resolution of the national-scale study is relatively crude. Therefore, on-installation ecosystem details are not captured.
3. The classification of ecosystem type on the installations is likely to be crude relative to the on-installation knowledge of local ecologists.
4. The forecasted change identifies the very long-term steady state of an area. It does not take into account the rate of change to that system, which is mediated by seed dispersal rates, longevity of mature trees, human system management initiatives, susceptibility to disease, and inter-species competition.
5. An entire line of research will need to be completed to understand how, when, and if ecosystems will actually shift according to changes in their suitable range. In the meantime, it is sufficient to understand that changing conditions are likely to put existing ecosystems in stress.

6.4.1 Canadian Model, Scenario A1b "Globalized Rapid Economic Growth-Bailey's Ecosystems"

Using the Bailey's ecosystem characterizations (Figures 21 to 23) we derived an evaluation of those installations that will experience an ecosystem shift. Of the 128 installations investigated, fully 96 (or 75%) changed from the climate ecosystem characteristic with which they started in the year 2000. An example of how suitable ranges for ecosystems are expected to migrate in the years 2000, 2050 and 2099 is presented in Figure 32. The bottom line is that installations are more likely to experience a major change than to remain static. Appendix A lists installations that shifted, and the climate regime to which they shifted.

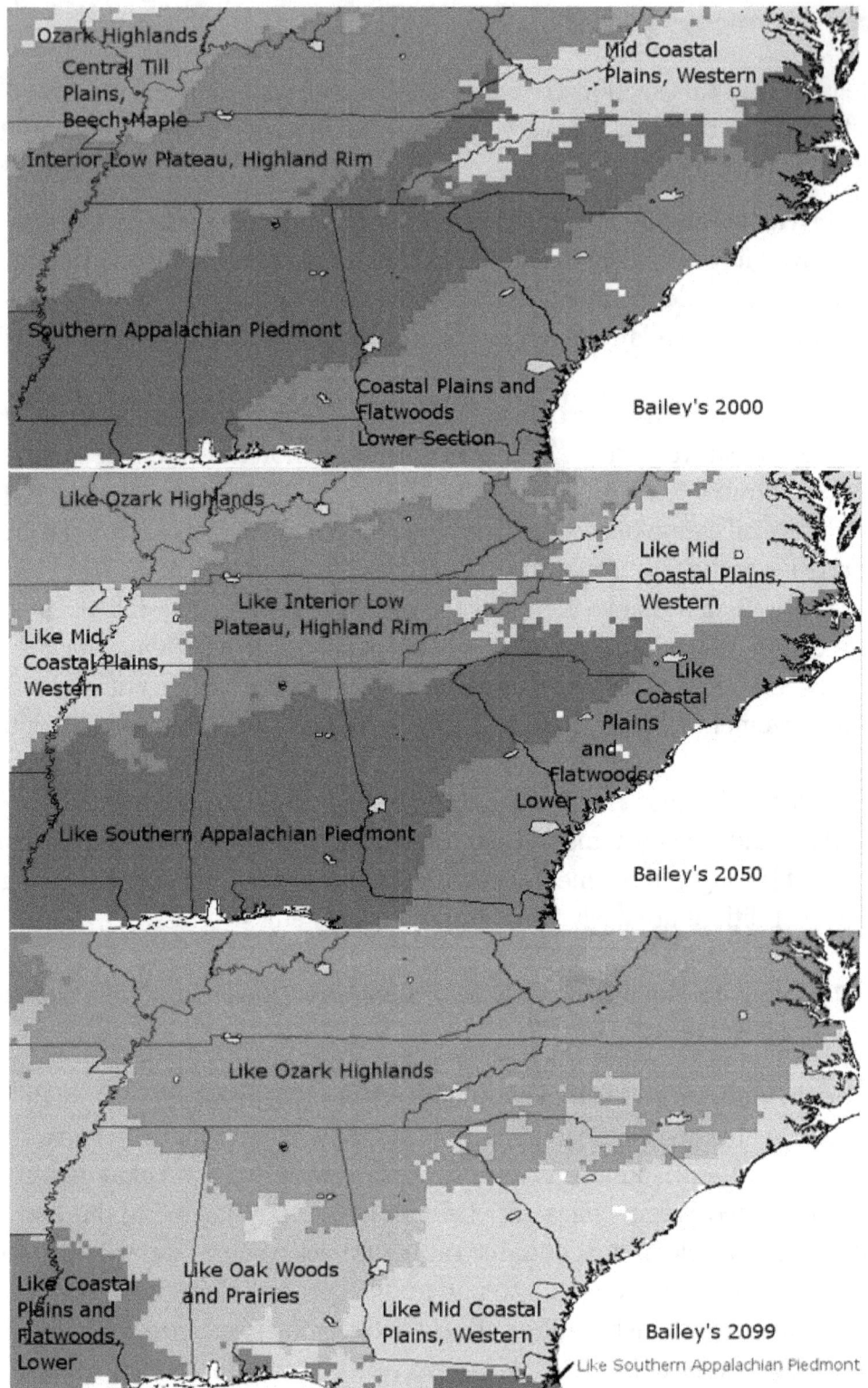

Figure 32. Change in "Predictive" version of Bailey's Ecosystems in the Southeastern States. A strong southern migration is evident. Climate change is often thought of as "warming." In this region, the migration is controlled by a major decrease in precipitation while temperatures change to only a minor degree.

6.4.2 Canadian Model, Scenario A1b "Globalized Rapid Economic Growth - Omernik's Ecosystems"

Similar to the Bailey's ecosystem change analysis, another was carried out using Omernik's Ecosystem characterization of the United States (a section of which is shown in Figure 33). In this analysis, 102 of the 128 installations studied (nearly 80%) shifted from their current climate-based ecosystem suitability range to another – indicating that the climate conditions that have brought on and supported the installations' current ecosystem regime will not exist at that installation in the future. Once more, some degree of change in the natural landscape is much more likely than stability during the next 100 years. Since there is a time lag between temperature and precipitation changes and responses in the ecosystem it needs to be clearly stated that these changes will slowly follow behind climate changes.

Installation land managers can expect a continuously changing landscape in both the near future and the long term horizon. Managing for preservation simply will not be an option in the future. This therefore implies that issues dealing with TES and invasive species will become increasingly problematic. Whole new areas of land management research must emerge to determine how the Army/DoD will change its management plans and how it will have to modify its current agreements with other Agencies (e.g., Forest Service and Fish and Wildlife Service) based on climate change dynamics. Again, Appendix A lists installations that shifted, and the climate regime to which they shifted.

6.4.3 Ecosystem changes based on Hargrove Data

We have previously introduced the work of Hargrove and Hoffman on ecosystem characterization. Although not a traditional ecosystem identification, we believe the technique has great potential for multiple applications. In this section, we introduce some of our results, so that their work may be seen in a balanced manner with others that deal specifically with climate change. In addition, the Hargrove/Hoffman approach proved so rich it warranted a separate sister report (Westervelt and Hargrove 2010) to deal with the implications of the data in much greater detail. The purpose of this report is to rank order climate impacts on Army installation.

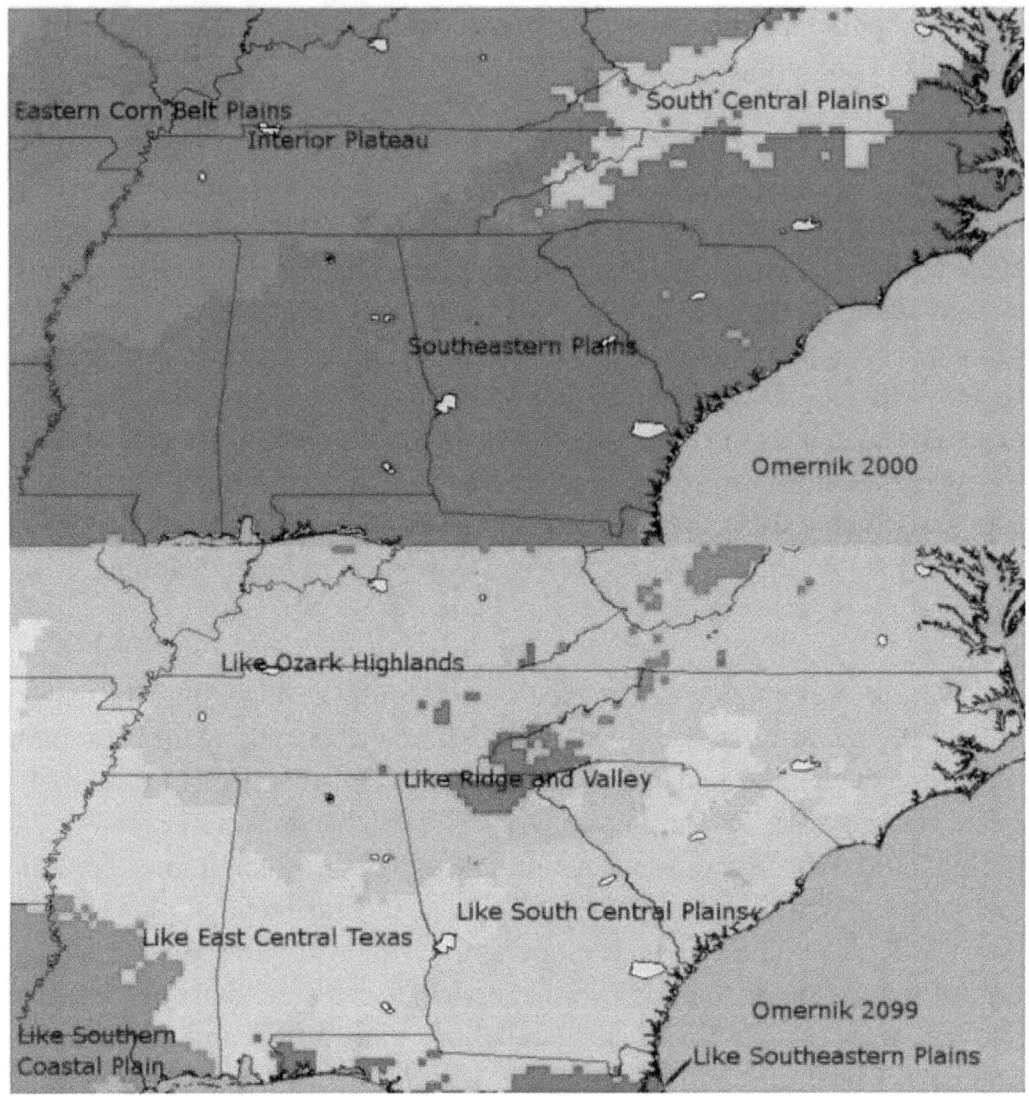

Figure 33. Migration of Omernik's Ecosystems in the Southeast United States (top) to the South and East (bottom). The same color table is used for both images.

Using the Hargrove data (which uses NCAR's PCM and the Hadley Centre's HADcm3), we were able to calculate the degree of change in an ecosystem's suitable range, which could not be done easily with Bailey's or Omernik's ecosystem characterizations (Appendix B). Table 15 lists the installations that, according to the Hargrove data and analysis, exhibit the greatest amount of change. The installations listed in Table 15 are distributed across the United States (Figure 34), but they correspond closely with many of our previous analyses. With the exception of Warrenton Training Center, Sunflower Army Ammunition Plant and Umatilla Chemical Depot, every installation listed in Table 15 has appeared elsewhere in this report.

Table 15. Percent of habitat unchanged by climate change.

Installation	Size (0.02 by 0.02) degree cells	PCM Model				Hadley Model			
		PCM B1 2050	PCM B1 2100	PCM A1 2050	PCM A1 2100	HAD B1 2050	HAD B1 2100	HAD A1 2050	HAD A1 2100
Fort Lee Military Reservation	66	36	50	50	27	50	0	0	0
Warrenton Training Center Military Reservation	130	47	25	64	18	31	18	22	0
Umatilla Chemical Depot (Closed)	88	59	55	27	23	24	20	24	0
Anniston Army Depot	90	70	53	70	20	11	11	11	0
Sunflower Army Ammunition Plant	35	83	83	83	0	0	0	0	0
Fort Bliss McGregor Range	11190	35	31	29	35	43	44	23	12
Pine Bluff Arsenal	288	89	83	87	3	0	0	0	0
Fort Sill Military Reservation	900	42	38	38	38	38	38	38	0
Fort Campbell	925	96	74	82	4	19	0	1	0
Fort Gordon	792	78	64	80	12	19	13	13	0
Key		80-100% unchanged	50-80% unchanged	0-50% unchanged					

Figure 34 shows an example of how these results can be interpreted. It examines the data from both the PMC and Hadley models and illustrates predicted changes at Fort Stewart, GA. The region is shown with the installation outlined in red. Fort Stewart currently resides primarily within the GAP ecoregion *Evergreen Plantations or Managed Pine*. The configuration of the currently existing nearby ecoregions is shown in the upper right insert of Figure 35. Compare that current situation insert with the predictions for the year 2050 (second row of inserts) for the PCM model, scenarios b1 and a1 and the Hadley model, scenarios b1 and a1 respectively. Relatively little of the suitable ecosystem range is predicted to change at Fort Stewart by the year 2050.

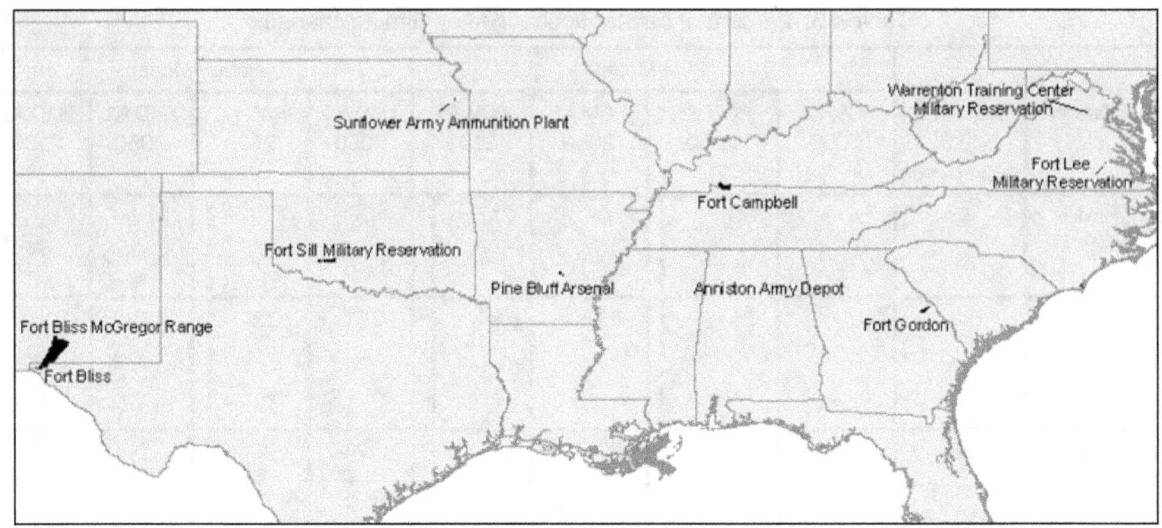

Figure 34. Installations showing the greatest ecological change based on the Hargove MGC data (Umitilla Chemical Depot is in the Northwest beyond the extent of this map).

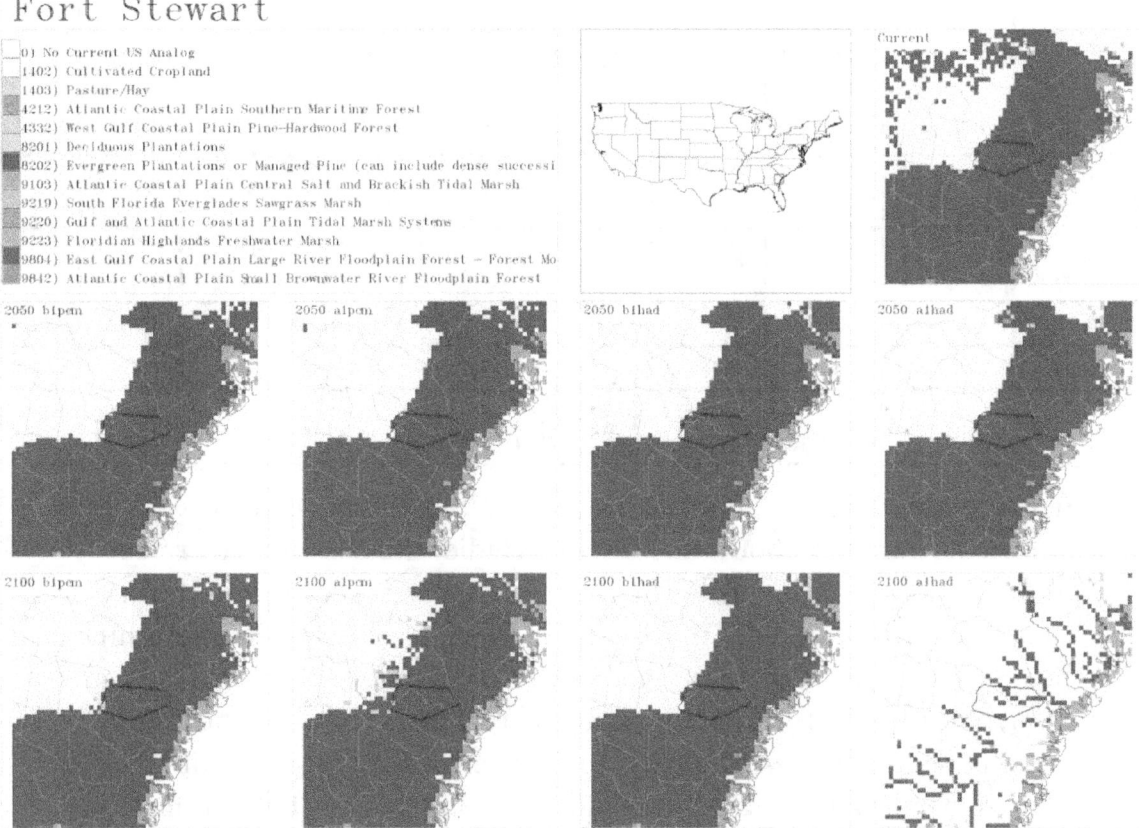

Source: http://earth.cecer.army.mil/LandSimModel/?q=node/116

Figure 35. More detailed MCI based analysis at Fort Stewart.

Next, compare the current situation insert with the predictions for the year 2100 (third row of inserts) for the PCM model, scenarios b1 and a1 and the Hadley model, scenarios b1 and a1 respectively. The PCM models suggest that Fort Stewart will be little changed. The Hadley model under the "sustainability" scenario (B1) shows little change in the *Evergreen Plantations or Managed Pine* distribution at Fort Stewart.

However, the Globalized Rapid Economic Growth scenario (A1) shows that the *Evergreen Plantations* type will retreat to remain only in the river valleys. Fort Stewart is projected to be completely covered by an ecosystem type without a current analogue. The land managers at Fort Stewart in this scenario could possibly be dealing with land management questions and issues for which there is currently no similar landscape. They could be managing their installation without historic guidance as there would be no example region from which they can take lessons or follow examples.

More detailed and extensive analysis types are offered in Volume II (Westervelt and Hargrove 2010) of this report. The most up-to-date version of these studies can be examined through the Internet at: http://earth.cecer.army.mil/LandSimModel/?q=node/116

6.5   Analysis of climate change effects on erosion

Table 16 lists the Tier 1 and 2 installations inside the CONUS. These installations are essential to consider due to their considerable size and great importance to the Army's mission. Table 17 lists those Tier 1 and 2 installations that ranked in either the High or Very High risk of Erosion due to climate change in 2099. Table 18 lists erosion risk of Tier 1 & 2 installations.

Three of the four Tier 1 and 2 installations with "very high risk" classifications (Fort Drum, Fort Indiantown Gap and Fort Dix) are located in the northeastern region of the United States – the area of the country projected to experience the greatest increase in precipitation intensity measured as a "Simple Daily Intensity Index" (SDII). The fourth "very high risk" installation is Fort Lewis – located on the other side of the county in western Washington. Spatial patterns of high risk installations are directly correlated to the straightforward nature of the precipitation intensity data. In fact, all of the Tier 1 and 2 installations classified as "very high" and "high" risk are concentrated in one of three clusters – the Pacific Northwest, the Upper Midwest and the Atlantic Northeast (Figure 36).

Table 16. Tier 1 & 2 Army Installations.

Installation Name	Tier	Installation Name	Tier
Fort Irwin	1	Fort Leonard Wood	2
Fort Polk	1	Fort Pickett	2
Fort Polk (Pelham Range)	1	Fort Sill	2
Fort Bragg	1	Camp Atterbury	2
Fort Bliss	1	Camp Blanding	2
Fort Bliss McGregor Range)	1	Fort Knox	2
Fort Lewis / YTC	1	Fort Ripley	2
Fort Hood	1	Fort Rucker	2
Fort Benning	1	Fort Chaffee	2
Fort Drum	1	Fort A.P. Hill	2
Fort Campbell	1	Fort Indiantown Gap	2
Fort Stewart & HAAF	1	Gowen Field Training Area	2
Fort Carson	1	Camp Grayling	2
Fort Carson (Pinyon Canyon)	1	Camp Bullis	2
Camp Riley	1	Fort Dix	2
Camp Shelby	2	Hunter-Liggett	2
Fort McCoy	2	Fort Jackson	2

Table 17. Tier 1 & 2 installations with
high or very high risk of erosion.

Installation Name	
Fort Dix	Fort A.P. Hill
Fort Indiantown Gap	Fort McCoy
Fort Lewis	Fort Knox
Fort Drum	Fort Pickett
Camp Atterbury	Fort Campbell
Yakima Firing Center	Camp Grayling

There are some notable connections between shift in suitable ecosystem ranges and projected erosion potential. Land cover and vegetation is particularly important to erosion, though we did not include that variable in this initial report. In the event that a limited number of specific installations are identified as requiring a deeper look into erosion potential, it will be possible to follow the method outlined in Zhang, et al. (2010) in which the authors used temporal downscaling to estimate CLIGEN input parameters to generate daily weather series representing future climates. Appendix C includes maps supporting this erosion analysis.

Table 18. Erosion risk on Tier 1 & 2 installations.

Key Installations	Rank
Fort Dix Military Reservation	
Fort Indiantown Gap Military Reservation	
Fort Lewis Military Reservation	
Fort Drum	
Camp Atterbury Military Reservation	
Yakima Firing Center	
Fort A.P. Hill Military Reservation	
Fort McCoy	
Fort Knox	
Fort Pickett Military Reservation	
Fort Campbell	
Camp Grayling Military Reservation	
Fort Bragg Military Reservation	
Hunter-Liggett Military Reservation	
Camp Ripley	
Camp Bullis	
Fort Jackson	
Fort Stewart	
Fort Leonard Wood Military Reservation	
Fort Hood	
Fort Benning Military Reservation	
Fort Chaffee	
Gowen Field Training Area	
Fort Irwin	
Fort Polk (Pelham Range)	
Fort Rucker Military Reservation	
Fort Sill Military Reservation	
Fort Polk Military Reservation	
Camp Shelby	
Fort Riley Military Reservation	
Camp Blanding Joint Training Center	
Fort Bliss	
Fort Bliss McGregor Range	
Fort Carson (Pinyon Canyon)	
Fort Carson Military Reservation	
	= "Very High Risk"
	= "High Risk"
	= "Moderate Risk"
	= "Low Risk"

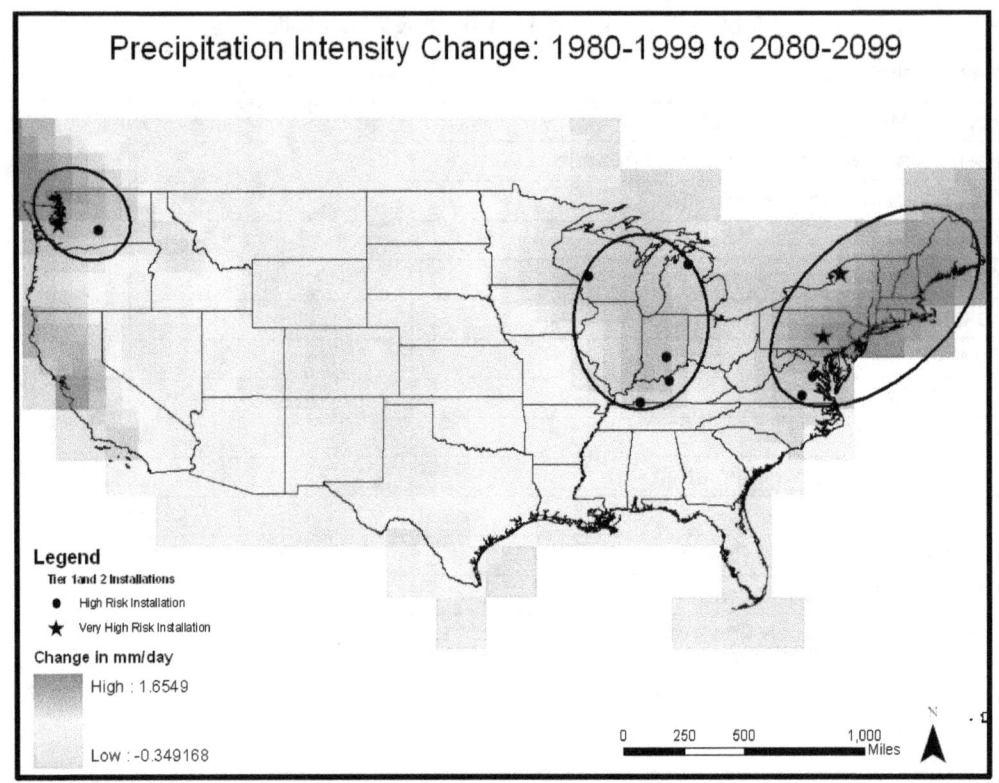

Figure 36. All of the highest risk Tier 1 and 2 installations are located in one of three clusters.

They then used those CLIGEN outputs to calculate specific data such as the *EI* necessary for an accurate R-factor to include in a projected RUSLE calculation. An analysis of this depth requires a specified location – thus the requirement to identify only a few installations. If calculated, this projected R-factor could be combined with other known and projected scenarios that take other aspects of RUSLE into account and thus provide a *much* more comprehensive idea of future erosion potential. Various scenarios could be tested to understand the potential impacts of alternative land management practices and the consequences of a possible shift in ecosystem suitability.

Ultimately, this erosion risk analysis provides a rank-order of Army installations that allows us to better understand one of the problems exacerbated by climate change by the end of the 21st Century. Among the most useful ways to apply this analysis is to use the relative risks between installations to better target specific installations for detailed study that will provide more concrete and comprehensive projections regarding future erosion risk.

6.5.1 References supporting the erosion analysis section

Gao, C., J. Zhu, Y. Hosen, J. Zhou, D. Wang, L. Wang, and Y. Dou. 2005. Effects of extreme rainfall on the export of nutrients from agricultural land. *Acta Geographica Sinica*. 60:991–97 (in Chinese with English abstract). Cited in: W. Wei, L. Chen, and B. Fu. 2009. Effects of rainfall change on water erosion processes in terrestrial ecosystems: a review. *Progress in Physical Geography*. 33(3):313.

Kharin, Viatcheslav V., Francis W. Zwiers, Xuebin Zhang, and Gabriele C. Hegerl. 2007. Changes in temperature and precipitation extremes in the IPCC ensemble of global coupled model simulations. *Journal of Climate*. Vol 20.

Meehl, Gerald A., Julie M. Arblaster, and Claudia Tebaldi. 2005. Understanding future patterns of increased precipitation intensity in climate model simulations. *Geophysical Research Letters*. Vol 32.

Meehl, G.A., T.F. Stocker, W.D. Collins, P. Friedlingstein, A.T. Gaye, J.M. Gregory, A. Kitoh, R. Knutti, J.M. Murphy, A. Noda, S.C.B. Raper, I.G. Watterson, A.J. Weaver and Z.-C. Zhao. 2007. Global Climate Projections. In: Climate Change 2007: The Physical Science Basis. Contribution of Working Group I to the Fourth Assessment Report of the Intergovernmental Panel on Climate Change [Solomon, S., D. Qin, M. Manning, Z. Chen, M. Marquis, K.B. Averyt, M. Tignor and H.L. Miller (eds.)]. Cambridge University Press, Cambridge, United Kingdom and New York, NY, USA.

Wei W, Chen L, and L. Y. Fu B. 2009. Effects of rainfall change on water erosion processes in terrestrial ecosystems: A review. Progress in Physical Geography. 33(3):307-318.

Zhang Y-G., M.A. Nearing, X.-C. Zhang, Y. Xie, and H. Wei, 2010. Projected rainfall erosivity changes under climate change from multimodel and multiscenario projections in Northeast China. Journal of Hydrology. 384:97-106.

6.6 Analysis of climate change effects on TES

Since the projection of individual TES habitat changes is well beyond the scope of this preliminary report, this section will rely on making observations of what projected climate changes might imply for TES at military installations. We suggest that this is basic groundwork for more detailed and in depth studies relating to climate change impacts on TES.

The Bailey's and Omernik's ecosystem migration results indicate that well over three-quarters of CONUS Army installations are likely to experience a shift in suitable ecosystem range due to climate change. From the complete Hargrove/ GAP-based results shown in Appendix B, Section 6.4.3, roughly 65% of the 133 installations are in the *Red: 50% or greater change in ecosystem* category from at least one of the eight different models used. So the evidence is overwhelming that there will be major changes from the present situation. Since any species depends on its current community composition for habitat and survival resources, the change of an entire ecosystem, even over hundreds of years, will have many implications for species survival. When suitable range for a species shifts away from its current location, that species experiences considerable stress. Such stress is even more significant for TES that are already struggling. General consequences that can reliably be based on these types of considerations include:

- The likelihood of current TES surviving will be greatly decreased regardless of time or money expended.
- As pockets of habitat locally shrink and existing species experience further stress, new TES will emerge.
- As suitable species ranges shift, some species will dramatically gain habitat, but *not locally*.
- Locally, large numbers of species will be more challenged for access to traditional resources. Therefore, the number of new TES candidates will increase dramatically if not overwhelmingly.
- The Army's current policy of managing for preservation is likely to become outdated.
- New ecological situations not currently in existence will emerge on Army lands; therefore:
  - The Army/DoD will have to modify its current agreements with other Agencies based on climate change dynamics.
  - The cost to manage TES (as well as Army lands in general) will greatly increase and there is no reason to believe it will cease rising since ecosystems will really only have begun migration/shift at the end of the available projections in 2099.
  - Whole new areas of land management research will emerge to understand specifically how ecosystems will react to shifted climate conditions and how the Army/DoD will change its TES management plans (as well as many other management plans).

Our focus TES for this report are the Red Cockaded Woodpecker and the Gopher Tortoise (GT). Our first step is to extract from general habitat descriptions those characteristics to which we can relate ecosystem and climate change qualities (Appendix D). Both species were chosen for this report because they are of great concern to the Army and their natural ranges are within the most climatically changed regions as previously described.

Red Cockaded Woodpecker (*Picoides borealis*) viability requires social groups of 25 or more breeding pairs in large areas (more than 2500 acres) of mature (minimum 70 year old) open mid-story pine forests that burn over every couple to a half dozen years (to maintain their open character). They thrived in the warm humid climate of the pre-colonial Southeast United States. The RCW tend to establish definitive territories and do not migrate very far from their breeding nests even when freshly matured and thus most prone to be looking for new nesting territories. They cannot endure many manmade industrial or urban areas.*

Our analysis suggests that the region the RCW inhabit is projected to become considerably drier and a little cooler. A drier climate suggests that the required fires may become more frequent. A cooler climate suggests that their habitat will shift south. In fact, the data in Figures 37 and 38 indicate that the current suitable conditions in which they currently live will migrate to Florida by 2099.

This does not imply that their habitat itself will shift in the same timeframe, but simply that the conditions that support the existing ecosystem will have shifted. More sophisticated study will be necessary how, when and if ecosystems will actually change. It is the character of the animal to not migrate; it is unlikely to be able to transfer to a new habitat. It has been shown that, even if encouraged, appropriate habitat will not be filled with RCW (Walters 1991). Thus, it can be expected that if their current habitat disappears in the future; the RCW will disappear from all of its current range. Thus, at Forts Benning, McPherson, Gordon, Bragg, Stewart, Rucker, Gillem, Camp MacKall, Hunter Army Airfield, and Anniston Army Depot, it is possible that the current RCW programs will experience stress due to events beyond their control. Habitats may continue to exist, but it is unlikely their numbers will expand, even in the unlikely case that new habitat is made available to them.

There are several critical climate related concerns for the habitat of the Gopher Tortoise (*Gopherus polyphemus*). They were formerly common in upland ecosystems throughout the southeastern United States, occupying well-drained sandy soils in partly canopied habitats with an abundant of herbaceous food at the floor level.

---

* Habitat information based on data in NatureServe web pages: <u>RCW</u> and <u>GT</u>.

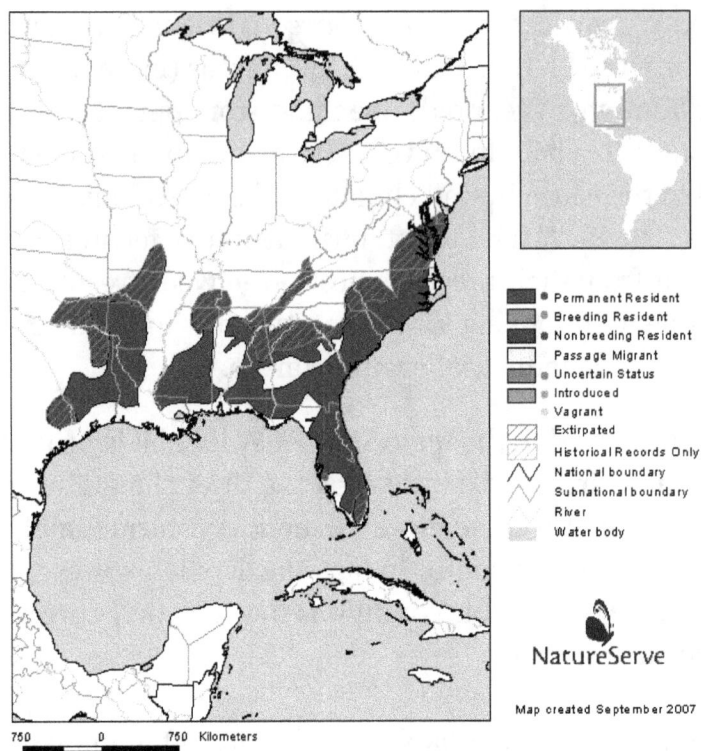

Figure 37. NatureServe map of red cockaded woodpecker distribution.

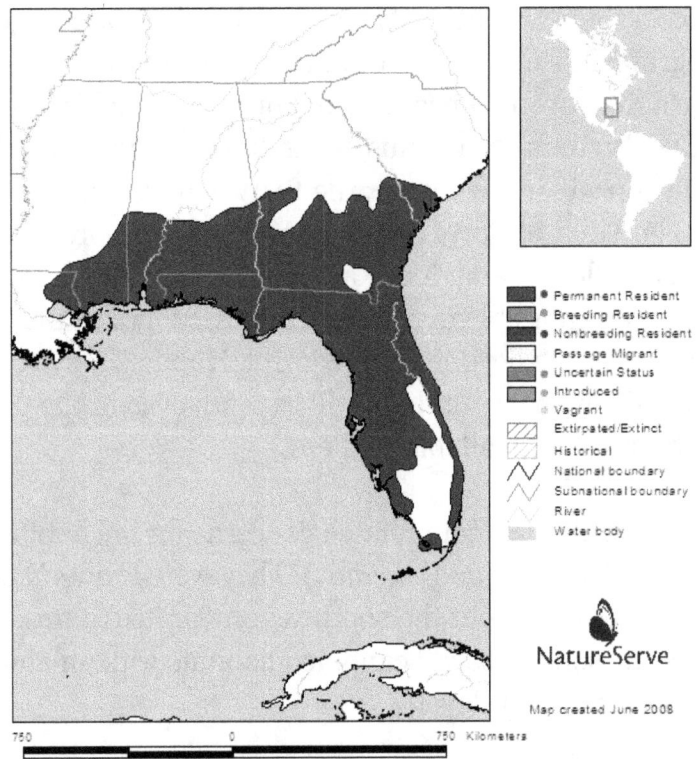

Figure 38. NatureServe map of gopher tortoise distribution.

To manage the openness, prescribed burning with a frequency of 2 to 7 years is recommended. They can withstand relatively high body temperatures (no higher than 43.9 °C [111 °F]) and prefer warmer periods. During colder periods, they tend to hibernate, so the further south one goes, the greater the phase of the year in which they are active. The highest densities can be found seeking the sun light to raise their body temperatures in their grassy, open-canopied sites. They do not do well when the larger areas of habitat are fragmented by human development (housing, agriculture, commercial forestry), denser forests (without the herbaceous floor) and major roads. They rapidly desiccate when deprived of their burrow. Individuals remain in their home range for many years, while switching burrows or digging new ones that they use to protect themselves from predators and climate extremes. They will migrate a short distance (about .5km) and will abandon stressed environments (e.g., occasionally subject to multi-year droughts).

As with the Red Cockaded Woodpecker, the data in Figures 37 and 38 indicate that the current suitable conditions in which the Gopher Tortoise currently live will migrate to Florida by 2099. The Southeastern states will become much drier. The drier climate may favor more frequent natural fires, but the herbaceous vegetation they depend on is likely to change character to a form they do not favor. The slightly cooler climate will slightly increase their time of inactivity, thus decreasing (slightly) opportunities for procreation. Though they do not migrate very far, a slow climate change will allow them time to move toward the south. It is possible with the climate shift that new unoccupied acceptable habitat will appear. They are capable of migrating, so if ecosystems do eventually shift to meet changed suitable conditions it is possible that the gopher tortoise will shift with it. Suitable range conditions for the tortoise are projected to disappear from Forts Benning, McPherson, Gordon, Bragg, Stewart, Rucker, Gillem, Camp MacKall, Hunter Army Airfield, and Anniston Army Depot. Gopher Tortoise's prognosis for expansion in Florida is a little better than the RCW, largely because they seem more capable of migration.

6.7 Analysis of climate change effects on invasive species

As with the issue of TES, this section relies on observations of what projected climate changes imply for invasive species at military installations.

The most important single observation is that the climate conditions supporting ecosystems will likely change in a manner that currently supports different ecosystems. This is likely to result in a series of ecological consequences that may lead to the establishment of the different ecosystem.

How and how fast that process proceeds is unknown and could take decades to centuries depending on many factors associated with ecosystem resiliency, persistence, and change thresholds. Part of that process is likely to result in the loss of some current species and an increase in population of new species, including invasives because:

- Invasives from similar ecosystems are likely to become established because global travel is so pervasive.
- Consequently, the difficulty with treating invasives in a traditional manner will be great.
- Financial and labor resources to deal with invasives will be strained if not broken.

The entire disciplinary subject has to change dramatically at its core. Since the climate basis for the "current situation" will change, we will be challenged to redefine the concept of invasives. For example, in the new ecosystem, will the residual old community members be called TES, or are they now to be known as invasive species? Obviously such a fundamental question will require an entirely new research.

However, we suggest the following considerations for invasive species responding to climate change:

- Invasives will become more common no matter how they are defined
- Invasives will be derived from sources that are:
  o Local residents, now decreasing in number due to a new climate for which they are unfit.
  o Near local sources that are migrating to keep up with the changing climate; that is, they represent the new ecosystem range – so are they really invasives?
  o Exotic sources from distant regions. As is the case now, species will use transportation (shipping and air travel in particular) to invade new habitats. If the volume of transportation increases, so will the number of exotic invasives.

When invasives arrive from distant regions several important events occur. Under climate change, not only can invasives survive in their adopted location, but many local species are experiencing increased survival threats and are thus weakened in their ability to compete against the new arrivals. Under this option, the creation of an entirely new ecosystem is possible in the long term. Further study will be necessary to understand if, when, and how ecosystems can be expected to change. When invasives populate an area, the resulting community tends to become less complicated, so new ecosystems are likely to have fewer members. When ecosystems become simpler, the traditional roles of the old ecosystem members are replaced or eliminated entirely. Thus, concerns such as degree of top soil erosion, wild fire vulnerability, and flash flood potential will be greatly enhanced in some locales.

7　Summary and Recommendations

As with other agencies, the effects of climate change are expected to impact CONUS military installations. Climate change has the potential to affect these major concerns at most CONUS installations:

- precipitation amounts
- temperature values
- ecosystem stability, type and/or traits
- erosion characteristics
- the management of TES
- the emergence and increase of invasive species.

The purpose of this report was to preliminarily evaluate using basic data and readily available information to begin providing answers to these questions, issues, and concerns in a scientific manner.

7.1　Precipitation

- Amount of change at Army installations range from -9 to + 10.6 mm/day
- Most significantly changed installations tend to clump into specific regions
- Installations in the Southeast will experience the greatest drying trend
- Installations along the mid to southern California coast will experience the greatest increase in rainfall
- Installations that regularly appear as being highly impacted under different models and different model scenarios are:
  - for drying conditions:
    * Fort Campbell
    * Fort Knox
    * Milan Arsenal and Wildlife Management Area
    * Camp Atterbury Military Reservation
    * Redstone Arsenal
    * Pine Bluff Arsenal
  - for increasingly wet conditions:
    * Hunter-Liggett Military Reservation
    * Presidio of Monterey
    * Camp Roberts Military Reservation.

7.2   Temperature

- Amount of change at Army installations range from -9.6 to +9.3 °C/month (−17.3 to 16.7 °F/month)
- Mostly highly changed installations tend to clump into specific regions
- Installations in the Southeast will experience the greatest decrease in temperatures.
- Installations in the Texas, New Mexico and Colorado region will experience the greatest increase in temperatures.
- Installations that regularly appear as being highly impacted under different models and different model scenarios are:
  - for decreasing temperature conditions (less drastic in the Australian Model):
    * Fort Rucker Military Reservation
    * Fort Benning Military Reservation
    * Fort Bragg Military Reservation
    * Fort Gillem Heliport
    * Fort Stewart
    * Fort McPherson
    * Camp MacKall Military Reservation
    * Hunter Army Airfield
    * Fort Gordon
    * Anniston Army Depot
  - for increasing temperature conditions:
    * Fort Wolters
    * Fort Hood
    * Camp Swift N. G. Facility
    * Buckley Air National Guard AF Base
    * Camp Bullis.

7.3   Ecosystems

It is very important to interpret this ecosystem analysis with the following caveats:

1. The Hadley and PCM models were chosen to represent relative extremes in GCM forecasts. The Canadian and Australian models were chosen to represent moderate forecasts. Similarly, the A1 and B1 gas-emission scenarios provide relative extremes in greenhouse gas emission rates over the 21st century.

2. Compared with the size of installations, the resolution of the national-scale study is relatively crude. Therefore, on-installation ecosystem details are not captured.

3. The classification of ecosystem type on the installations is likely to be crude relative to the on-installation knowledge of local ecologists.

4. The forecasted change identifies the very long-term steady state of an area. It does not take into account the rate of change to that system, which is mediated by seed dispersal rates, longevity of mature trees, human system management initiatives, susceptibility to disease, and inter-species competition.

5. An entire line of research will need to be completed to understand how, when, and if ecosystems will actually shift according to changes in their suitable range. In the meantime, it is sufficient to understand that changing conditions are likely to put existing ecosystems in stress.

Ecosystem change predictions were modeled under three different ecosystem definitions (Bailey's, Omernik's, and the USGS 2010 GAP analysis) and under three different climate models using two different scenarios of the direction of future growth. In all cases:

- Changes will occur all across CONUS and will often be major changes (as measured by percent of ecosystem on an installation unchanged by 2099).

- The percent of Army installations that will change from their current climate ecosystem range to a new one ranged from a low of 50% to a high of 80%. Thus a strained or fluctuating ecosystem by 2099 will be the normal situation for Army land managers.

- Changes will often be to new climate ecosystem characteristics that do not currently exist; therefore, no current ecosystem will exist on which installation land managers will be able to model their management activities.

- The mostly highly altered ecosystems will be at:
  o Fort Lee Military Reservation
  o Warrenton Training Center Military Reservation
  o Umatilla Chemical Depot (Closed)
  o Anniston Army Depot
  o Sunflower Army Ammunition Plant
  o Fort Bliss McGregor Range
  o Pine Bluff Arsenal
  o Fort Sill Military Reservation
  o Fort Campbell
  o Fort Gordon.

7.4   Erosion

While changes in average precipitation measures are important, erosion is predominantly dependent on the number and character of extreme rainfall events. The IPCC and others predict not only an upward trend in total precipitation, but also an increasing bias toward more extreme participation events. Thus, we can expect rainfall intensity to increase in many regions through the year 2099 and therefore can expect greater potential for erosion. We used precipitation intensity projections derived from a multi-model ensemble that calculated the projected difference in precipitation intensity between the end of the 20th century and the end of the 21st. The data allowed us an approximate look at how future precipitation intensity might influence the risk of erosion at Army Installations across the CONUS. We found that the following Tier 1 and 2 installations rank highest in potential risk of erosion due to climate change by 2099:

- Fort Dix
- Fort Indiantown Gap
- Fort Lewis
- Fort Drum
- Camp Atterbury
- Yakima Firing Center
- Fort A.P. Hill
- Fort McCoy
- Fort Knox
- Fort Pickett
- Fort Campbell
- Camp Grayling.

7.5   TES

- The likelihood of currently indentified TES surviving will be greatly decreased no matter how much money or time is expended. It is possible that their local habitat may be stressed, shifted, or may disappear entirely.
- As pockets of habitat locally shrink, many new TES will emerge.
- As ecosystems shift, some species will dramatically gain habitat, but *not locally*.
- Locally large numbers of species will be more challenged for access to traditional resources. Therefore, locally, the number of new TES candidates will increase dramatically if not overwhelmingly.

- Managing for preservation simply will not be an option.
- The potential for new ecosystems that are not currently in existence will emerge on Army lands, therefore whole new areas of land management research will emerge dealing with how the Army/DoD will change its TES management plans (as well as many other management plans).
- The Army/DoD will have to modify its current agreements with other Agencies based on Climate change dynamics.
- The cost to manage TES (as well as Army lands in general) will greatly increase and there is no reason to believe it will not continue to rise since ecosystems will almost certainly not stop migrating at the end of our available projections in 2099.

7.6   Invasive Species

Many of the conclusions for the TES analysis apply to noxious invasive species as well. In addition to those concerns above, and specifically for the concern of invasive species:

- As ecosystems (or climate ecosystem characteristics) shift, some species will gain habitat.
- Invasives will become more common no matter how they are defined.
- Invasives from similar ecosystems are likely to become established because global travel is so pervasive. Invasives will be derived from sources that are
   o Local residents, now decreasing in number due to a new climate for which they are unfit.
   o Near local sources that are migrating to keep up with the changing climate; that is, they represent the new ecosystem – so are they really invasives?
   o Exotic sources from distant regions. As is the case now, species will use transportation (shipping and air travel in particular) to invade new habitats. If the volume of transportation increases, so will the number of exotic invasives.
      *   Thus the difficulty with treating invasives in a traditional manner will be great,
      *   Financial and labor resources to deal with invasives will be strained if not broken

7.7   Installations at greatest risk due to climate change

It is apparent from the lists presented in the precipitation, temperature, ecosystem and erosion analyses above that certain installations appear multiple times for high probability of modification by 2099 due to climate change. Since this report was developed based on the research of three independent investigations by three independent researchers using many different climate model predictions and several different scenarios, there can be little question that when these Army installations appear repeatedly, the best available data is indicating the changes will be real and dramatic. All the various research efforts that supported this work examined well over one hundred installations, but the lists provided in this chapter included only the top 10 changed installations, i.e., less than 8% of all the locations evaluated. Consequently, when an installation appears multiple times in multiple lists, significant changes are highly likely, considering that:

- Fort Campbell appears in three lists.
- Fort Knox appears in two lists.
- Camp Atterbury Military Reservation appears in two lists.
- Pine Bluff Arsenal appears in two lists.
- Fort Gordon appears in two lists.
- Anniston Army Depot appears in two lists.

Figure 39 shows the distribution of these Army installations.

After so many variations in analyses it is significant that these six installations are distributed over such a small section of the United States. It is apparent that other installations not on this list, but in the region (named in Figure 39), are likely to share the fate and problems that these six will experience.

An installation not included in this list will not escape climate change. As mentioned above, anywhere from 65–80% of the installations are expected to experience change from their current ecosystem identification. That is no small matter. It implies many subordinate changes in the character and species that reside in those locations. Throughout this report and particularly in this section we have highlighted the greatest changes; though military land managers can expect climate change challenges almost anywhere in CONUS.

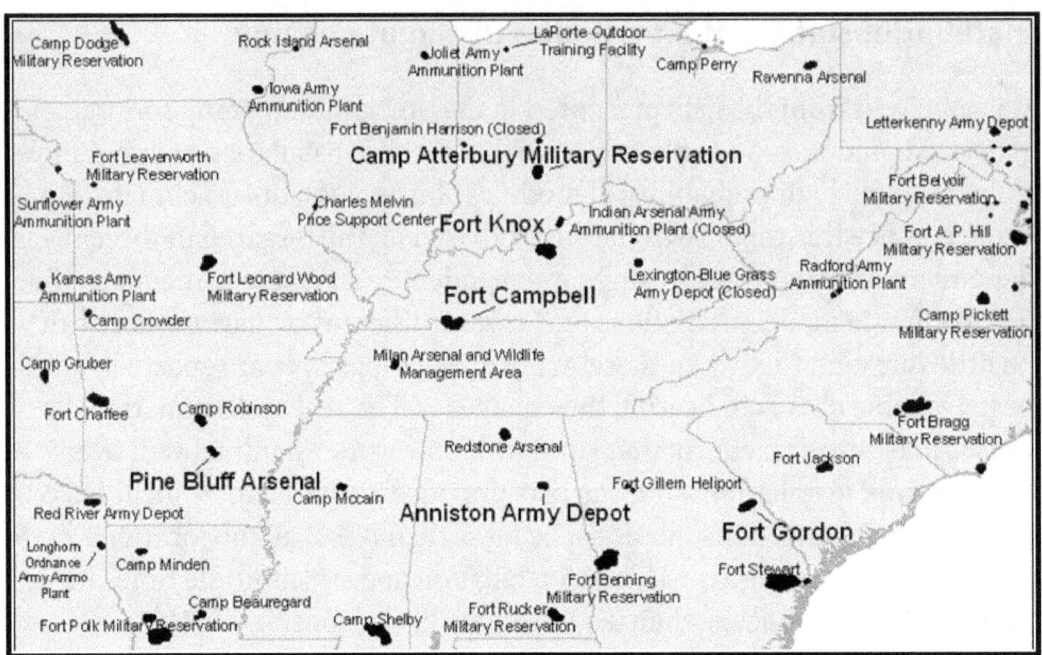

Figure 39. Installations most often highly impacted by climate change concerns from multiple sources (large bold black font) and others in the same region (smaller blue font).

As a concise summary of what this report has determined, we will answer a few common questions:

- Will climate change affect Army installations in the next 100 years?
  - Yes
- In the next 50 years?
  - Probably
- Will Army installations not on the coastline and therefore not subject to sea level rise be affected?
  - Yes
- Will major changes affect a few, many, or most installations?
  - Most
- Will the Army need to change any of its current procedures to respond to these changes?
  - Yes
- Are these changes expected to have marginal or major impacts to Army land managers?
  - Major
- Are impacts dispersed or concentrated?
  - Climate change affects all installations.
  - The worst impacts tend to group into regions.

o Multiple high impacts will occur in the region of the Mississippi and Ohio River Valleys as well as most of the Southeastern United States (not necessarily including Florida).

- What are the major areas of concern for Army land managers?
  o Ecosystem change
  o Erosion
  o Greater TES issues
  o Greater invasive species issues.

7.8 Recommendations

7.8.1 For Army trainers

Climate change implies that installation missions may have to change. For example, if your primary "tropical" training installation is likely to become much drier, you are likely to look elsewhere for an appropriate training facility. As climate changes, the supplies required to support your field personnel will have to change too. In a hotter climate, for example, you will require more drinking water. But where are you going to get it if the local supply is limited?

If erosion increases on a tracked vehicle training facility (as would have been the case for the old mission at Fort Knox, a multi-impacted location), can it still sustain the training exercises it was established to support?

The implications of climate change need to be investigated in relation to installation missions at the DoD, Army and Regional levels.

7.8.2 For Army land managers

What installation land managers can expect in the near future and over the long term is a continuously changing landscape. Managing for preservation simply will not be an option. This therefore implies that issues dealing with TES and invasive species will not only persist, but will worsen. It also implies that some issues may become less of a concern as conditions shift. Costs to manage Army lands in the traditional manner will skyrocket. New suitable ranges for ecosystems will emerge for which no current analogue or model exists. Traditional land management techniques will no longer work. So a new area of land management research must emerge to determine how the Army/DoD will change its management plans and how it will have to modify its current agreements with other Agencies (e.g., Forest Service and Fish and Wildlife Service) based on climate change dynamics.

7.8.3 For additional research

Since TES are already an important issue at many important installations and this work indicates the problem will balloon in size, shouldn't we now be developing research and a complete Army program to deal with the question before it becomes unmanageable?

Noxious invasive species will become an issue in which even the definition of the area will be difficult. We need to develop a parallel management and research program in this area to robustly handle the problems climate change will present.

We mentioned that ecosystem change is likely to result in new systems. We need to know:

- If, how and when can we expect changes in character of the current ecosystems on military installations?
- How will we handle Army land with natural communities that have no current analogue?
- How will we deal with new problems created due to ecosystem change such as increased erosion and flash flooding due to a simpler ecosystem that does not provide adequate vegetative cover?

7.8.4 Final word

This work is a preliminary report. There is clearly much to which the Army must respond to continue to effectively carry on its military preparedness mission. It is recommended that an in depth research campaign be carried out in at least three contrasting regions of the United States to begin to deal in greater depth with climate changes the Army will experience.

References

Albergel, J., S. Nasri, J. M. Lamache`re. 2004. HYDROMED – Programme de recherche sur les lacs collinaires dans les zones semi- arides du pourtour mé diterrané en. Revue des Sciences de l'Eau 17 (2):133-151.

Bailey, Robert G. 1983. Delineation of Ecosystem Regions. Environmental Management 7(4):365-373.

———. 1995. Description of the ecoregions of the United States (2d ed.). Misc. Pub. No. 1391, Map scale 1:7,500,000. Washington, DC: US Department of Agriculture (USDA) Forest Service.

———. 1991. Design of Ecological Networks for Monitoring Global Change. Environmental Conservation 18(2):173-175.

———. 1988. Ecogeographic Analysis. USDA Forest Service: Misc Publication 1465.

———. 2004. Identifying Ecoregion Boundaries. Environmental Management 34(Supplement 1):S14-S26.

———. 1984. Testing an Ecosystem Regionalization. Journal of Environmental Management 19:239-248.

———. 1984. The Factor of Scale in Ecosystem Mapping. 1984. Environmental Management 9(4):271-276.

Boer, G. J. 1995: A hybrid moisture variable suitable for spectral GCMs. Research Activities in Atmospheric and Oceanic Modelling. Report No. 21, WMO/TD-No. 665. World Meteorological Organization, Geneva, Switzerland.

Cai, W., M. A. Collier, P. D. Durack, H. B. Gordon, A. C. Hirst, S. P. O'Farrell, and P. H. Whetton. 2003. The response of climate variability and mean state to climate change: Preliminary results from the CSIRO Mark 3 coupled model. CLIVAR Exchanges. 28:8-11.

Cai, W., G. Meyers, and G. Shi. 2005. Transmission of ENSO signals to the Indian Ocean. Geophys. Res. Let. In press.

Cai, W., G. Shi, Y. Li. 2005. Multidecadal fluctuations of winter rainfall over southwest Western Australia simulated in the CSIRO Mark 3 coupled model. Geophys. Res. Let. submitted.

Cai, W., H. Hendon, and G. Meyers. 2005. An Indian Ocean Diploe-like variability in the CSIRO Mark 3 climate model. J. Climate. In press.

Cai, W., M. A. Collier, H. B. Gordon, and L. J. Waterman. 2003. Strong ENSO variability and a super-ENSO pair in the CSIRO coupled climate model. Monthly Weather Review. 131:1189-1210.

Cai, W., M. J. McPhaden, M. A. Collier. 2004. Multidecadal fluctuations in the relationship between equatorial Pacific heat content anomalies and ENSO amplitude. Geophys. Res. Let. 31, L01201. DOI:10.1029/2003GL018714.

Carter, T. R., R. N. Jones, X. Lu, S. Bhadwal, C. Conde, L.O. Mearns, B.C. O'Neill, M. D. A. Rounsevell, and M. B. Zurek. 2007. New assessment methods and the characterisation of future conditions. Climate Change 2007: Impacts, Adaptation and Vulnerability. Contribution of Working Group II to the Fourth Assessment Report of the Intergovernmental Panel on Climate Change. Cambridge, United Kingdom (UK): Cambridge University Press. 133-171.

Chris Gordon, Claire Cooper, Catherine A. Senior, Helene Banks, Jonathan M Gregory, Timothy C Johns, John FB Mitchell, and Richard A Wood. 2000. The simulation of SST, sea ice extents and ocean heat transports in a version of the Hadley centre coupled model without flux adjustments. Climate Dynamics. 16:147-168. Bracknell, UK: Hadley Centre for Climate Prediction and Research.

Christensen, J. H., B. Hewitson, A. Busuioc, A. Chen, X. Gao, I. Held, R. Jones, R.K. Kolli, W.-T. Kwon, R. Laprise, V. Magaña Rueda, L. Mearns, C.G. Menéndez, J. Räisänen, A. Rinke, A. Sarr, and P. Whetton. 2007. Regional climate projections. in Climate Change 2007: The Physical Science Basis. Contribution of Working Group I to the Fourth Assessment Report of the Intergovernmental Panel on Climate Change. Cambridge, UK and New York, NY, USA: Cambridge University Press.

Commonwealth Scientific and Industrial Research Organisation (CSIRO). 2006. CSIRO climate model results used for international benchmarking studies. Web page, http://www.csiro.au/science/climate-model-results.html

Favis-Mortlock D, Boardman J. 1995. Nonlinear responses of soil erosion to climate change: A modelling study on the UK South Downs. Catena. 25(1-4):365-387.

Favis-Mortlock D., Guerra, A. 1999. The implications of general circulation model estimates of rainfall for future erosion: A case study from Brazil. Catena. 37(3-4):329-354.

Field J. P., D. D. Breshears, J. J. Whicker. 2009. Toward a more holistic perspective of soil erosion: Why aeolian research needs to explicitly consider fluvial processes and interactions. Aeolian Research. 1(1-2):9-17.

Field, C. B., L. D. Mortsch, M. Brklacich, D. L. Forbes, P. Kovacs, J. A. Patz, S. W. Running, and M. J. Scott. 2007. North America. Climate change 2007: Impacts, adaptation and vulnerability. Contribution of working group II to the fourth assessment report of the Intergovernmental Panel on Climate Change. Cambridge, UK: Cambridge University Press.

Gao, C., J. Zhu, Y. Hosen, J. Zhou, D. Wang, L. Wang, and Y. Dou. 2005. Effects of extreme rainfall on the export of nutrients from agricultural land. *Acta Geographica Sinica*. 60:991–97 (in Chinese with English abstract). Cited in: W. Wei, L. Chen, and B. Fu. 2009. Effects of rainfall change on water erosion processes in terrestrial ecosystems: a review. *Progress in Physical Geography*. 33(3):313.

Gerald A. Meehl, Julie M. Arblaster, and Claudia Tebaldi. 2005. Understanding future patterns of increased precipitation intensity in climate model simulations. *Geophysical Research Letters*. 32:1, DOI: 10.1029/2005GL023680.

Global Environmental and Climate Change Centre (GEC3), Environment Canada (Adaptation and Impact Research Section) and the Drought Research Initiative (DRI). 2010. Data Access Integration Portal. Global climate models daily data. Web page, http://loki.ouranos.ca/DAI/gcm-e.html

Gordon, Chris, Claire Cooper, Catherine A. Senior, Helene Banks, Jonathan M Gregory, Timothy C Johns, John F. B. Mitchell, and Richard A. Wood. 2000. The simulation of SST, sea ice extents and ocean heat transports in a version of the Hadley Centre Coupled Model without flux adjustments. Climate Dynamics. 16:147-168. Bracknell, UK: Hadley Centre for Climate Prediction and Research.

Gordon, H. B., L. D. Rotstayn, J. L. McGregor, M. R. Dix, E. A. Kowalczyk, S. P. O'Farrell, L. J. Waterman, A. C. Hirst, S. G. Wilson, M. A. Collier, I. G. Watterson, and T. I. Elliott. 2002. The CSIRO Mk3 climate system model [Electronic publication]. CSIRO Atmospheric Research technical paper No. 60. Aspendale: CSIRO Atmospheric Research, http://www.dar.csiro.au/publications/gordon_2002a.pdf

Hargrove, William W., and Forrest Hoffman. 2004. Potential of multivariate quantitative methods for delineation and visualization of ecoregions. Environmental Management. 34(1): S39-S60. DOI: 10.1007/s00267-003-1084-0. New York, NY: Springer.

Headquarters, US Army Corps of Engineers (HQUSACE). 2007. US Army Installation Floristic Inventory Database. Public Works Technical Bulletin (PWTB) 200-1-52, Washington, DC: HQUSACE.

Hibler, W. D. 1979. A dynamic thermodynamic sea ice model. J. Phys. Oceanogr. 9:815-846.

Intergovernmental Panel on Climate Change (IPCC). 2001. J. T. Houghton, Y. Ding, D. J. Griggs, M. Noguer, P. J. van der Linden, X. Dai, K. Maskell, and C.A. Johnson (eds.). *Climate Change 2001: The Scientific Basis. Contribution of Working Group I to the Third Assessment Report of the Intergovernmental Panel on Climate Change*. Cambridge, UK: Cambridge University Press. http://www.globalwarmingart.com/wiki/File:Global_Warming_Predictions_png

————. 2007. S. Solomon, D. Qin, M. Manning, Z. Chen, M. Marquis, K.B. Averyt, M. Tignor and H.L. Miller (eds.). 2007. *The Physical Science Basis, Contribution of Working Group I to the Fourth Assessment Report of the Intergovernmental Panel on Climate Change.* Cambridge, United Kingdom and New York, NY, USA: Cambridge University Press, http://ipcc-wg1.ucar.edu/wg1/wg1-report.html

————. 2007a. S. Solomon, D. Qin, M. Manning, Z. Chen, M. Marquis, K.B. Averyt, M. Tignor and H.L. Miller (eds.). Summary for Policymakers. In: *Climate Change 2007: The Physical Science Basis. Contribution of Working Group I to the Fourth Assessment Report of the Intergovernmental Panel on Climate Change.* Cambridge University Press, Cambridge, United Kingdom and New York, NY, USA, http://www.ipcc.ch/pdf/assessment-report/ar4/wg1/ar4-wg1-spm.pdf

————. 2010. The IPCC Data Distribution Centre. AR4 GCM Data. Web page, http://www.mad.zmaw.de/IPCC_DDC/html/SRES_AR4/index.html

Karl, T. R., and R. W. Knight. 1998. Secular trends of precipitation amount, frequency, and intensity in the United States. Bulletin of the American Meteorological Society. 79(2):231-241.

Kharin, Viatcheslav V., Francis W. Zwiers, Xuebin Zhang, and Gabriele C. Hegerl. 2007. Changes in temperature and precipitation extremes in the IPCC ensemble of global coupled model simulations. *Journal of Climate.* Vol 20.

Manabe, S., and R. T. Wetherald. 1967. Thermal equilibrium of the atmosphere with a given distribution of relative humidity. Journal of the Atmospheric Sciences. 24(3):241-259.

Maurer, E. P., L. Brekke, T. Pruitt, and P. B. Duffy. 2007. Fine-resolution climate projections enhance regional climate change impact studies. *Eos Trans. AGU.* 88(47):504.

Mearns, L. O., W. J. Gutowski, R. Jones, L.-Y. Leung, S. McGinnis, A. M. B. Nunes, and Y. Qian. 2009. A regional climate change assessment program for North America. *Eos Trans. AGU.* 90(36):311-312.

Meehl, Gerald A., Julie M. Arblaster, and Claudia Tebaldi. 2005. Understanding future patterns of increased precipitation intensity in climate model simulations. *Geophysical Research Letters.* Vol 32.

Meehl, G. A., C. Covey, T. Delworth, M. Latif, B. McAvaney, J. F. B. Mitchell, R. J. Stouffer, and K. E. Taylor. 2007. The WCRP CMIP3 multi-model dataset: A new era in climate change research, *Bulletin of the American Meteorological Society.* 88:1383-1394.

Meehl, G.A., T.F. Stocker, W.D. Collins, P. Friedlingstein, A.T. Gaye, J.M. Gregory, A. Kitoh, R. Knutti, J.M. Murphy, A. Noda, S.C.B. Raper, I.G. Watterson, A.J. Weaver and Z.-C. Zhao. 2007. Global Climate Projections. In: Climate Change 2007: The Physical Science Basis. Contribution of Working Group I to the Fourth Assessment Report of the Intergovernmental Panel on Climate Change [Solomon, S., D. Qin, M. Manning, Z. Chen, M. Marquis, K.B. Averyt, M. Tignor and H.L. Miller (eds.)]. Cambridge University Press, Cambridge, United Kingdom and New York, NY, USA.

National Aeronautics and Space Administration (NASA). 2011. Earth Observatory. Net primary productivity. Web page, http://earthobservatory.nasa.gov/GlobalMaps/view.php?d1=MOD17A2_M_PSN

Nearing, M. A., V. Jetten, C. Baffaut, O. Cerdan, A. Couturier, M. Hernandez, Y. Le Bissonnais, M. H. Nichols, J. P. Nunes, C. S. Renschler, V. Souchere, and K. Van Oost. 2005. Modeling response of soil erosion and runoff to changes in precipitation and cover. *Catena*.61(2-3):131-154.

Nunes J. P., J. Seixas, J. J. Keizer, and A. J. Ferreira. 2009. Sensitivity of runoff and soil erosion to climate change in two Mediterranean watersheds. Part II: Assessing impacts from changes in storm rainfall, soil moisture and vegetation cover. *Hydrological Processes*. 1220(March):1212- 1220.

Nunes P., N.R. Pacheco. 2008. Vulnerability of water resources, vegetation productivity and soil erosion to climate change in Mediterranean watersheds. Hydrological Processes. 3134(October 2007):3115- 3134.

Office of Assistant Chief of Staff for Installation Management (OACSIM). Army Environmental Division, Installation Services Directorate. 2009. Installation Summaries from the FY2007 Survey of Threatened and Endangered Species on Army Lands. San Antonio, TX: Headquarters, Installation Management Command (HQIMCOM), p 2.

Omernik, J. M. 1987. Ecoregions of the conterminous United States (map supplement): Annals of the Association of American Geographers, 77(1):118-125, http://www.jstor.org/stable/2569206?cookieSet=1

Omernik, James M., and Robert G. Bailey. 1997. distinguishing between watersheds and ecoregions. Journal of the American Water Resources Association. 33(5):935-949.

O'Neal, M., M. Nearing, R. Vining, J. Southworth, and R. Pfeifer. 2005. Climate change impacts on soil erosion in Midwest United States with changes in crop management. *Catena*. 61(2-3):165-184.

Pruski F., and M. Nearing. 2002. Climate-induced changes in erosion during the 21st century for eight US locations. Water Resources Research. 38(12).

Raclot, D., and J. Albergel. 2006. Runoff and water erosion modelling using WEPP on a Mediterranean cultivated catchment. Physics and Chemistry of the Earth, Parts A/B/C. 31(17):1039.

Renard, K. G., G. R. Foster, G. A. Weesies, D. K. McCool, and D. C. Yoder (coordinators). 1997. Predicting Soil Erosion by Water: A Guide to Conservation Planning With the Revised Universal Soil Loss Equation (RUSLE). Agricultural Handbook No. 703. Washington, DC: US Department of Agriculture (USDA), p 14.

Secretary of Defense. 2010. *Quadrennial Defense Review Report*. Washington DC: Department of Defense, http://www.defense.gov/qdr/QDR%20as%20of%2026JAN10%200700.pdf

Smith, Wade, Daniel Schultz, Daniel Whitford, Judith Barry, Daniel Uyesugi, Will Mitchell, and Barry Stamey. 2010. *Climate Change Planning for Military Installations: Findings and Implications*. Falls Church VA: Noblis.

Soil and Water Conservation Society (SWCS). 2003. Conservation Implications of Climate Change: Soil Erosion and Runoff from Cropland. Ankeny, IA: SWCS.

SWCS. Conservation Implications of Climate Change: Soil Erosion and Runoff from Cropland. *Soil and Water*. 2003:24.

US Climate Change Science Program and the Subcommittee on Global Change Research. 2008. Preliminary Review of Adaptation Options for Climate-Sensitive Ecosystems and Resources. Final Report, Synthesis and Assessment. Product 4.4.

US Department of Energy (USDOE). 2004. Parallel climate Model. Web page, http://www.cgd.ucar.edu/pcm/

US Department of the Interior (USDOI). 2011. Bailey's Ecoregions and Subregions of the United States, Puerto Rico, and the US Virgin Islands. Web page, http://www.nationalatlas.gov/mld/ecoregp.html

Verseghy, D. L., N. A. McFarlane, and M. Lazare. 1993. A Canadian Land Surface Scheme for GCMs:II. Vegetation model and coupled runs. Int. J. Climatol. 13:347-370.

Walters, J. R. 1991. Application of ecological principles to the management of endangered species: The case of the red-cockaded woodpecker. Annual Rev. Ecol. Syst. 22:505-523.

Watterson, I. G., 2005: The intensity of precipitation during extra-tropical cyclones in global warming simulations: a link to cyclone intensities? Tellus A. Accepted for publication.

Watterson, I. G., and M. R. Dix. 2005. Effective sensitivity and heat capacity in the response of climate models to greenhouse gas and aerosol forcings. Q. J. Roy. Met. Soc., 131:259-280.

Weatherly, J.W., and C. M. Bitz. 2001. Natural and anthropogenic climate change in the arctic. 12th Symposium on Global Change and Climate Variations, 15-18 January 2001, Albuquerque, Boston, MA: American Meteorological Society.

Wei W, Chen L, and L. Y. Fu B. 2009. Responses of water erosion to rainfall extremes and vegetation types in a loess semiarid hilly area, NW China. Hydrological Processes. 1791(May):1780-1791.

————. 2009. Effects of rainfall change on water erosion processes in terrestrial ecosystems: A review. Progress in Physical Geography. 33(3):307-318.

Westervelt, James, and Michael Hargrove. 2010. Anticipating climate change impacts on Army installations: Volume 2. Champaign, IL: Engineer Research and Development Center-Construction Engineering Research Laboratory (ERDC-CERL). Submitted, November 2010.

Yang D, S. Kanae, T. Oki, T. Koike, and K. Musiake. 2003. Global potential soil erosion with reference to land use and climate changes. Hydrological Processes. 17(14):2913-2928.

Zhang X., and W. Liu. 2005. Simulating potential response of hydrology, soil erosion, and crop productivity to climate change in Changwu tableland region on the Loess Plateau of China. *Agricultural and Forest Meteorology*. 131(3-4):127-142.

Zhang X., Liu W., Z. Li, and F. Zheng. 2009. Simulating site-specific impacts of climate change on soil erosion and surface hydrology in southern Loess Plateau of China. *Catena*. 79(3):237-242.

Zhang X., and M. Nearing. 2005. Impact of climate change on soil erosion, runoff, and wheat productivity in central Oklahoma. *Catena*. 61(2-3):185-195.

Zhang X. 2007. A comparison of explicit and implicit spatial downscaling of GCM output for soil erosion and crop production assessments. Climatic Change. 84(3-4):337-363.

Zhang Y-G., M.A. Nearing, X.-C. Zhang, Y. Xie, and H. Wei, 2010. Projected rainfall erosivity changes under climate change from multimodel and multiscenario projections in Northeast China. Journal of Hydrology. 384:97-106.

Zhang, G. J., and N. A. McFarlane. 1995. Sensitivity of climate simulations to the parameterization of cumulus convection in the CCC-GCM. Atmos.-Ocean. 3:407-446.

Zhou, Z., Z. Shangguan, and D. Zhao. 2006.Modeling vegetation coverage and soil erosion in the Loess Plateau Area of China. *Ecological Modelling*. 198(1-2):263-268.

Acronyms and Abbreviations

Term	Definition
ACE	Army Corps of Engineers
AVHRR	Advanced Very High Resolution Radiometer
CEERD	US Army Corps of Engineers, Engineer Research and Development Center
CEQ	Council on Environmental Quality
CERL	Construction Engineering Research Laboratory
CONUS	Continental United States
CRREL	Cold Regions Research and Engineering Laboratory
CSIRO	Commonwealth Scientific and Industrial Research Organisation
DRI	Drought Research Initiative
EOS	Earth Observing System
ERDC	Engineer Research and Development Center
GAP	USGS Gap Analysis Program
GCM	Global Climate Model
GEC3	Global Environmental and Climate Change Centre
GHG	Greenhouse Gas
IPCC	Intergovernmental Panel on Climate Change
IPPC	Intergovernmental Panel on Climate Change
LANDFIRE	Landscape Fire and Resource Management Planning Tools
MCI	
MGC	Multivariate Geographic Clustering
MODIS	Moderate Resolution Imaging Spectroradiometer
NASA	National Aeronautics and Space Administration
NEPA	National Environmental Policy Act
NPV	net primary vegetation
PCM	Parallel Climate Model
PCMDI	Program for Climate Model Diagnosis and Intercomparison
QDR	Quadrennial Defense Review
SERDP	Strategic Environmental Research and Development Program
SI	Systeme Internationale
SRES	The Special Report on Emissions Scenarios
TES	threatened and endangered species
URL	Universal Resource Locator
US	United States
USA	United States of America
USDA	US Department of Agriculture
USEPA	US Environmental Protection Agency
USGS	US Geological Survey
WCRP	World Climate Research Programme
WWW	World Wide Web

Appendix A: Data Used To Generate Climate and Ecosystem Change Evaluations in Sections 6.2.1 to 6.4.2

Table A1. Scenario parameters.

Installation Name	Canadian Scenario A1b Prcp 2000 mm/day	Canadian Scenario A1b Prcp 2099 mm/day	Canadian Scenario A1b Prcp 20-99 Change mm/day	Canadian Scenario A1b Temp 2000 C_deg	Canadian Scenario A1b Temp 2099 C_deg	Canadian Scenario A1b Temp 20-99 Change C_deg	Canadian Scenario B1 Prcp 2000 mm/day	Canadian Scenario B1 Prcp 2099 mm/day	Canadian Scenario B1 Prcp 20-99 Change mm/day	Canadian Scenario B1 Temp 2000 C_deg	Canadian Scenario B1 Temp 2099 C_deg	Canadian Scenario B1 Temp 20-99 Change C_deg	Australian Scenario A1b Prcp 2000 mm/day	Australian Scenario A1b Prcp 2099 mm/day C_deg	Australian Scenario A1b Prcp 20-99 Change mm/day C_deg	Australian Scenario A1b Temp 2000 C_deg	Australian Scenario A1b Temp 2099 C_deg	Australian Scenario A1b Temp 20-99 Change C_deg
Aberdeen Proving Ground Military Reservation	6.0	2.9	-3.1	5.8	2.4	-3.4	6.0	2.5	-3.5	5.9	3.2	-2.7	2.1	4.4	2.3	0.2	2.0	1.8
Anniston Army Depot	6.3	3.4	-2.9	14.0	5.5	-8.6	6.3	2.8	-3.5	14.0	9.7	-4.4	7.3	2.7	-4.6	7.6	9.1	1.5
Arlington National Cemetery	5.5	2.6	-2.9	6.4	2.7	-3.7	5.5	2.1	-3.4	6.4	3.8	-2.6	2.1	3.7	1.6	0.9	2.3	1.4
Army Chemical Center	6.0	2.9	-3.1	5.8	2.4	-3.4	6.0	2.5	-3.5	5.9	3.2	-2.7	2.1	4.4	2.3	0.2	2.0	1.8
Army Reserve Outdoor Training Area	0.8	1.3	0.6	-5.7	-4.3	1.4	0.8	0.7	0.0	-5.8	-1.5	4.3	0.6	0.9	0.3	-5.5	-6.5	-1.0
Army Training Area	0.5	1.0	0.5	-5.4	-4.4	1.0	0.5	0.5	0.0	-5.4	-1.1	4.3	0.4	0.6	0.2	-5.7	-5.0	0.8
Badger Army Ammunition Plant	1.5	2.1	0.6	-7.5	-5.8	1.7	1.5	1.7	0.2	-7.9	-2.3	5.6	1.0	1.1	0.1	-9.0	-7.1	1.9
Bearmouth National Guard Training Area	1.5	1.0	-0.5	-9.1	-6.8	2.3	1.5	1.1	-0.4	-9.5	-1.2	8.3	0.6	0.7	0.2	-9.3	-5.1	4.2
Belle Mead General Depot	7.1	3.6	-3.5	2.3	1.8	-0.5	7.1	3.3	-3.8	2.4	1.3	-1.1	2.0	5.1	3.1	-2.1	0.4	2.5
Blossom Point Field Test Facility	5.3	2.7	-2.6	7.7	3.1	-4.5	5.3	2.1	-3.2	7.7	4.5	-3.2	2.2	3.7	1.5	1.6	3.3	1.6
Buckeye National Guard Target Range	0.1	0.2	0.1	8.9	13.2	4.3	0.1	1.0	0.9	8.3	12.8	4.5	0.6	1.3	0.6	9.4	14.6	5.1
Buckley Air National Guard AF Base	0.4	0.6	0.2	-4.1	1.3	5.4	0.4	0.3	-0.1	-3.7	3.9	7.6	0.2	0.4	0.1	-2.2	3.2	5.4
Camden Test Annex	5.0	4.0	-0.9	-4.8	-2.3	2.5	5.0	4.4	-0.5	-4.9	-3.5	1.5	2.1	3.4	1.4	-7.7	-4.5	3.2
Camp Adair Military Reservation	13.9	11.9	-2.0	3.3	4.8	1.5	13.9	14.0	0.1	3.5	7.2	3.7	5.2	8.6	3.4	2.6	5.6	3.0
Camp Atterbury Military Reservation	9.6	3.2	-6.3	0.7	-1.2	-2.0	9.6	3.7	-5.9	0.6	3.9	3.3	2.7	3.8	1.2	-2.5	-1.5	1.1
Camp Bullis	0.9	0.1	-0.7	8.4	11.5	3.1	0.9	0.6	-0.3	8.6	16.0	7.3	2.3	0.6	-1.8	7.3	13.0	5.7
Camp Dodge Military Reservation	1.1	1.7	0.5	-5.6	-4.6	1.0	1.1	0.9	-0.3	-5.7	-0.9	4.8	0.9	1.0	0.2	-5.5	-7.2	-1.7
Camp Grayling Military Reservation	2.8	3.1	0.3	-7.2	-5.1	2.1	2.8	3.0	0.1	-7.3	-5.0	2.3	1.4	1.7	0.3	-9.2	-5.3	3.9
Camp Johnson	2.4	2.4	0.0	-5.4	-2.3	3.1	2.4	1.9	-0.4	-5.3	-6.6	-1.3	1.1	2.3	1.2	-10.1	-4.8	5.3
Camp Joseph T. Robinson	6.2	2.1	-4.1	4.7	6.2	1.5	6.2	0.8	-5.4	5.0	9.5	4.5	4.0	2.3	-1.7	5.3	7.3	2.0
Camp MacKall Military Reservation	3.9	3.1	-0.8	13.8	5.0	-8.8	3.9	1.8	-2.1	14.0	7.9	-6.0	4.3	3.5	-0.8	6.7	8.1	1.4

Installation Name	Australian Scenario A1b Temp 2000 C_deg	Australian Scenario A1b Temp 2099 C_deg	Australian Scenario A1b Temp 20-99 Change C_deg	Australian Scenario A1b Prcp 2000 mm/day	Australian Scenario A1b Prcp 2099 mm/day	Australian Scenario A1b Prcp 20-99 Change mm/day	Canadian Scenario B1 Temp 2000 C_deg	Canadian Scenario B1 Temp 2099 C_deg	Canadian Scenario B1 Temp 20-99 Change C_deg	Canadian Scenario B1 Prcp 2000 mm/day	Canadian Scenario B1 Prcp 2099 mm/day	Canadian Scenario B1 Prcp 20-99 Change mm/day	Canadian Scenario A1b Temp 2000 C_deg	Canadian Scenario A1b Temp 2099 C_deg	Canadian Scenario A1b Temp 20-99 Change C_deg	Canadian Scenario A1b Prcp 2000 mm/day	Canadian Scenario A1b Prcp 2099 mm/day	Canadian Scenario A1b Prcp 20-99 Change mm/day
Camp Parks Military Reservation	7.4	11.1	3.8	1.7	10.2	8.5	7.7	11.9	4.1	2.4	7.8	5.4	7.8	9.8	2.0	2.4	1.9	-0.5
Camp Riley Military Reservation	5.3	7.6	2.3	7.4	8.3	0.8	6.0	9.7	3.7	15.0	14.0	-0.9	5.6	7.3	1.6	15.0	12.9	-2.1
Camp Roberts Military Reservation	7.8	11.7	3.8	1.3	8.8	7.5	7.5	12.5	5.0	1.0	6.5	5.5	7.4	10.1	2.7	1.0	1.1	0.1
Camp Swift N. G. Facility	7.8	13.3	5.6	3.4	0.9	-2.5	9.0	16.7	7.7	1.5	0.9	-0.6	8.3	12.0	3.8	1.5	0.4	-1.1
Camp Williams	-4.7	-0.9	3.8	0.6	4.4	3.8	-5.8	1.5	7.3	1.5	1.9	0.5	-6.3	-2.4	3.9	1.5	1.0	-0.5
Camp Williams	-10.3	-7.7	2.6	1.1	0.8	-0.3	-9.1	-3.4	5.7	1.8	1.9	0.1	-8.8	-7.1	1.7	1.8	2.1	0.4
Charles Melvin Price Support Center	-0.2	-1.1	-1.0	1.5	3.6	2.1	-1.2	5.2	6.4	3.8	1.3	-2.5	-1.1	0.9	2.1	3.8	2.3	-1.5
Cornhusker Army Ammunition Plant	-6.0	-4.7	1.3	0.4	0.6	0.3	-5.5	-0.9	4.6	0.5	0.4	-0.1	-5.5	-4.5	1.0	0.5	1.0	0.5
Custer Reserve Forces Training Area	-5.1	-2.3	2.8	1.5	3.1	1.6	-2.7	0.2	2.9	4.1	3.6	-0.6	-2.4	-1.9	0.5	4.1	3.3	-0.8
Dugway Proving Grounds	-2.9	1.3	4.2	0.2	1.2	1.0	-4.0	3.2	7.2	0.4	0.6	0.1	-4.3	-0.5	3.8	0.4	0.3	-0.1
Edgewood Arsenal	0.2	2.0	1.8	2.1	4.4	2.3	5.9	3.2	-2.7	6.0	2.5	-3.5	5.8	2.4	-3.4	6.0	2.9	-3.1
Florence Military Reservation	10.4	15.0	4.6	0.9	1.3	0.5	8.5	13.2	4.6	0.1	1.3	1.2	9.3	13.4	4.1	0.1	0.3	0.2
Fort A. P. Hill Military Reservation	1.5	3.2	1.7	2.5	3.4	0.9	8.0	4.2	-3.8	4.9	2.0	-2.9	8.0	2.4	-5.5	4.9	2.8	-2.1
Fort Belvoir Military Reservation	0.8	2.3	1.5	2.1	3.7	1.6	6.4	3.7	-2.6	5.4	2.1	-3.3	6.4	2.5	-3.8	5.4	2.6	-2.8
Fort Benning Military Reservation	10.6	11.6	0.9	6.3	3.3	-3.0	17.1	11.9	-5.1	4.4	3.2	-1.2	17.0	7.7	-9.3	4.4	3.2	-1.3
Fort Bliss	4.4	10.7	6.2	0.5	0.3	-0.2	5.3	9.1	3.8	0.1	0.3	0.2	5.3	8.4	3.1	0.1	0.1	0.1
Fort Bliss McGregor Range	2.9	9.0	6.1	0.7	0.6	-0.1	4.1	7.7	3.6	0.1	0.4	0.3	4.0	7.0	3.0	0.1	0.2	0.1
Fort Bragg Military Reservation	6.5	7.9	1.4	4.3	3.4	-0.8	13.8	7.6	-6.1	3.8	1.8	-2.0	13.6	4.7	-8.9	3.8	3.1	-0.7
Fort Campbell	2.0	4.2	2.2	5.0	3.2	-1.8	5.2	6.8	1.6	10.9	1.8	-9.0	5.2	2.0	-3.2	10.9	3.0	-7.8
Fort Carson Military Reservation	-0.8	4.8	5.6	0.2	0.4	0.3	-1.8	5.0	6.8	0.3	0.2	-0.1	-1.9	2.1	4.0	0.3	0.3	0.0
Fort Carson Military Reservation	-1.7	3.6	5.3	0.2	0.3	0.1	-3.2	4.0	7.1	0.3	0.2	-0.1	-3.3	1.7	5.1	0.3	0.4	0.1
Fort Dix Military Reservation	-0.8	1.7	2.5	2.1	5.4	3.3	4.5	2.5	-2.0	7.0	3.1	-3.9	4.4	2.5	-1.9	7.0	3.5	-3.4

Installation Name	Prop 2000 mm/day Canadian Scenario A1b	Prop 2099 mm/day Canadian Scenario A1b	Prop 20-99 Change mm/day Canadian Scenario A1b	Temp 2000 C_deg Canadian Scenario A1b	Temp 2099 C_deg Canadian Scenario A1b	Temp 20-99 Change C_deg Canadian Scenario A1b	Prop 2000 mm/day Canadian Scenario B1	Prop 2099 mm/day Canadian Scenario B1	Prop 20-99 Change mm/day Canadian Scenario B1	Temp 2000 C_deg Canadian Scenario B1	Temp 2099 C_deg Canadian Scenario B1	Temp 20-99 Change C_deg Canadian Scenario B1	Prop 2000 mm/day Australian Scenario A1b	Prop 2099 mm/day C_deg Australian Scenario A1b	Prop 20-99 Change mm/day C_deg Australian Scenario A1b	Temp 2000 C_deg Australian Scenario A1b	Temp 2099 C_deg Australian Scenario A1b	Temp 20-99 Change C_deg Australian Scenario A1b
Fort Drum	3.8	3.4	-0.4	-6.2	-3.4	2.8	3.8	3.3	-0.5	-6.3	-5.8	0.6	1.8	2.8	1.1	-9.2	-5.6	3.7
Fort Ethan Allen Military Reservation	3.0	3.0	0.0	-6.0	-3.0	3.0	3.0	2.4	-0.6	-5.9	-7.1	-1.2	1.3	2.8	1.6	-10.8	-5.5	5.2
Fort Eustis Military Reservation	4.6	3.4	-1.2	11.8	4.2	-7.6	4.6	2.1	-2.5	11.9	6.3	-5.6	3.2	3.7	0.5	4.1	6.4	2.4
Fort George G. Meade	5.7	2.7	-3.0	5.9	2.2	-3.7	5.7	2.2	-3.5	6.0	3.3	-2.6	2.1	3.9	1.8	0.4	1.9	1.5
Fort Gillem Heliport	5.1	3.5	-1.6	14.4	5.5	-8.9	5.1	3.3	-1.8	14.5	9.8	-4.7	6.5	3.5	-3.0	8.2	9.3	1.2
Fort Gordon	3.7	3.9	0.2	14.8	6.0	-8.8	3.7	4.1	0.4	14.9	10.0	-4.9	6.0	4.5	-1.5	8.7	9.7	1.0
Fort Hood	1.3	0.4	-0.9	5.7	10.2	4.5	1.3	0.5	-0.9	6.7	15.1	8.3	2.6	0.6	-2.0	5.8	11.6	5.8
Fort Huachuca	0.1	0.5	0.4	6.5	10.4	3.9	0.1	1.2	1.0	6.1	10.1	4.1	0.8	0.9	0.1	7.0	11.9	4.8
Fort Irwin	0.2	0.3	0.1	2.8	6.3	3.6	0.2	2.6	2.4	2.5	8.4	5.9	0.4	4.1	3.7	3.8	8.7	4.9
Fort Jackson	3.8	3.7	-0.1	14.4	6.0	-8.4	3.8	2.9	-0.9	14.6	9.4	-5.2	4.9	3.9	-1.0	8.1	9.1	1.0
Fort Knox	11.1	3.1	-8.1	4.8	1.3	-3.6	11.1	3.2	-7.9	4.7	6.8	2.1	3.8	3.0	-0.8	0.9	2.7	1.8
Fort Leavenworth Military Reservation	1.3	1.2	-0.1	-1.5	-0.6	0.9	1.3	0.4	-0.9	-1.7	4.0	5.7	0.8	1.4	0.6	-0.8	-2.9	-2.1
Fort Lee Military Reservation	4.4	3.2	-1.2	11.5	3.7	-7.8	4.4	2.0	-2.5	11.5	5.9	-5.6	3.1	3.4	0.3	3.7	6.0	2.3
Fort Leonard Wood Military Reservation	3.5	1.6	-1.9	0.1	1.5	1.4	3.5	0.6	-2.8	0.4	5.2	4.8	1.6	2.4	0.9	0.7	0.5	-0.3
Fort Lewis Military Reservation	8.6	6.8	-1.8	4.0	5.7	1.7	8.6	7.2	-1.4	4.3	8.4	4.1	4.2	4.2	-0.1	3.9	6.1	2.1
Fort MacArthur	0.5	0.6	0.1	11.6	14.8	3.2	0.5	5.2	4.8	11.4	16.3	4.9	1.4	8.0	6.6	12.0	16.0	4.0
Fort McCoy	1.7	1.9	0.3	-9.3	-7.6	1.7	1.7	1.8	0.1	-9.7	-3.8	5.8	1.1	0.7	-0.4	-10.8	-8.0	2.8
Fort McPherson	5.2	3.4	-1.8	14.1	5.3	-8.8	5.2	3.1	-2.0	14.2	9.6	-4.6	6.4	3.3	-3.1	7.9	9.1	1.2
Fort Monmouth Military Reservation	7.2	3.7	-3.5	4.0	3.1	-0.9	7.2	3.2	-4.0	4.0	2.8	-1.2	2.1	5.4	3.3	-0.8	1.9	2.8
Fort Monroe Military Reservation	4.5	3.4	-1.1	11.9	4.4	-7.5	4.5	2.0	-2.4	12.0	6.5	-5.5	3.2	3.8	0.6	4.3	6.6	2.3
Fort Polk Military Reservation	4.8	2.0	-2.9	10.2	10.0	-0.3	4.8	2.1	-2.7	10.9	13.2	2.3	9.7	2.3	-7.4	8.7	11.2	2.5
Fort Riley Military Reservation	0.6	1.0	0.4	-2.7	-1.6	1.1	0.6	0.1	-0.5	-2.7	3.7	6.4	0.4	1.0	0.6	-1.9	-3.3	-1.4

Installation Name	Canadian Scenario A1b Prop 2000 mm/day	Canadian Scenario A1b Prop 2099 mm/day	Canadian Scenario A1b Prop 20-99 Change mm/day	Canadian Scenario A1b Temp 2000 C.deg	Canadian Scenario A1b Temp 2099 C.deg	Canadian Scenario A1b Temp 20-99 Change C.deg	Canadian Scenario B1 Prop 2000 mm/day	Canadian Scenario B1 Prop 2099 mm/day	Canadian Scenario B1 Prop 20-99 Change mm/day	Canadian Scenario B1 Temp 2000 C.deg	Canadian Scenario B1 Temp 2099 C.deg	Canadian Scenario B1 Temp 20-99 Change C.deg	Australian Scenario A1b Prop 2000 mm/day	Australian Scenario A1b Prop 2099 mm/day C.deg	Australian Scenario A1b Prop 20-99 Change mm/day C.deg	Australian Scenario A1b Temp 2000 C.deg	Australian Scenario A1b Temp 2099 C.deg	Australian Scenario A1b Temp 20-99 Change C.deg
Fort Ritchie Raven Rock Site	5.8	2.9	-2.9	3.0	0.6	-2.3	5.8	2.7	-3.1	3.0	1.5	-1.5	2.1	3.6	1.5	-2.0	-0.4	1.6
Fort Rucker Military Reservation	3.8	3.5	-0.3	18.7	9.1	-9.6	3.8	3.5	-0.3	18.7	13.0	-5.7	7.7	4.0	-3.7	12.3	12.6	0.2
Fort Sill Military Reservation	0.7	0.3	-0.4	0.7	5.8	5.0	0.7	0.2	-0.6	1.3	10.2	9.0	1.0	0.7	-0.3	2.7	7.1	4.4
Fort Stewart	2.8	4.1	1.3	18.2	9.3	-8.9	2.8	4.2	1.4	18.3	13.0	-5.3	4.8	5.4	0.6	13.3	13.2	-0.1
Fort William H. Harrison Military Reservation	1.1	0.9	-0.2	-10.2	-8.0	2.2	1.1	0.9	-0.2	-10.9	-1.5	9.3	0.5	0.7	0.2	-10.1	-6.0	4.1
Fort Wolters	1.1	0.2	-0.8	3.6	8.3	4.7	1.1	0.3	-0.8	4.4	13.4	9.0	1.8	0.4	-1.4	3.8	9.8	6.0
Globecom Radio Receiving Station	5.7	2.8	-2.9	6.8	2.7	-4.1	5.7	2.2	-3.5	6.8	3.9	-2.9	2.2	4.0	1.7	1.0	2.5	1.5
Greencastle Military Reservation	5.4	2.8	-2.6	3.0	1.1	-1.9	5.4	2.6	-2.8	3.0	1.8	-1.2	1.9	3.3	1.3	-1.8	-0.1	1.6
Hunter Army Airfield	2.7	3.8	1.2	17.5	8.7	-8.8	2.7	3.9	1.2	17.7	12.4	-5.2	4.4	4.8	0.4	12.5	12.5	-0.1
Hunter-Liggett Military Reservation	1.1	1.3	0.1	6.9	9.6	2.7	1.1	7.6	6.5	6.9	12.0	5.0	1.6	10.4	8.8	7.3	11.1	3.8
Hunter-Liggett Military Reservation	1.7	1.8	0.0	6.4	8.9	2.5	1.7	9.4	7.7	6.4	11.2	4.8	1.9	12.6	10.6	6.6	10.3	3.8
Iowa Army Ammunition Plant	2.4	2.5	0.1	-5.1	-2.5	2.6	2.4	1.8	-0.6	-4.8	1.6	6.4	1.1	2.5	1.5	-4.1	-6.5	-2.4
Joliet Army Ammunition Plant	3.6	2.9	-0.6	-3.6	-2.5	1.1	3.6	2.7	-0.8	-3.9	0.7	4.6	1.4	4.1	2.7	-5.1	-4.6	0.5
Joliet Army Ammunition Plant	3.2	2.8	-0.4	-3.7	-2.5	1.3	3.2	2.6	-0.6	-4.1	0.8	4.9	1.3	3.8	2.5	-5.1	-4.6	0.5
Kansas Army Ammunition Plant	1.7	0.9	-0.8	-0.2	1.9	2.1	1.7	0.3	-1.4	0.6	5.7	5.2	0.8	1.1	0.3	1.6	1.9	0.3
Kearney Rifle Range	0.4	0.8	0.4	-4.5	-3.5	1.0	0.4	0.3	-0.1	-4.5	0.3	4.8	0.3	0.6	0.2	-5.1	-3.4	1.7
Lake City Army Ammunition Plant	1.6	1.4	-0.2	-1.4	-0.4	1.0	1.6	0.5	-1.1	-1.5	4.1	5.6	0.9	1.7	0.8	-0.6	-2.6	-2.0
LaPorte Outdoor Training Facility	5.4	3.3	-2.1	-2.0	-2.0	0.0	5.4	3.3	-2.1	-2.1	0.6	2.7	1.6	4.8	3.1	-4.5	-3.2	1.3
Letterkenny Army Depot	5.2	2.8	-2.4	1.9	0.6	-1.3	5.2	2.7	-2.6	1.9	1.1	-0.8	1.8	3.0	1.3	-2.7	-0.9	1.8
Longhorn Ordnance Army Ammo Plant	3.9	1.4	-2.5	8.8	9.5	0.7	3.9	1.3	-2.5	9.0	13.1	4.2	5.0	1.5	-3.5	7.3	10.7	3.4
Los Alamitos Armed Forces Reserve Center	0.6	0.7	0.1	12.1	15.5	3.4	0.6	6.1	5.5	11.8	16.9	5.1	1.7	9.2	7.6	12.7	16.7	4.0
Louisiana Ordnance Plant	5.0	1.9	-3.1	8.6	9.0	0.4	5.0	1.6	-3.3	8.7	12.4	3.7	6.6	2.0	-4.5	7.4	10.1	2.7

Installation Name	Canadian Scenario A1b Prep 2000 mm/day	Canadian Scenario A1b Prep 2099 mm/day	Canadian Scenario A1b Prep 2020-99 Change mm/day	Canadian Scenario A1b Temp 2000 C deg	Canadian Scenario A1b Temp 2099 C deg	Canadian Scenario A1b Temp 2020-99 Change C deg	Canadian Scenario B1 Prep 2000 mm/day	Canadian Scenario B1 Prep 2099 mm/day	Canadian Scenario B1 Prep 2020-99 Change mm/day	Canadian Scenario B1 Temp 2000 C deg	Canadian Scenario B1 Temp 2099 C deg	Canadian Scenario B1 Temp 2020-99 Change C deg	Australian Scenario A1b Prep 2000 mm/day	Australian Scenario A1b Prep 2099 mm/day C deg	Australian Scenario A1b Prep 2020-99 Change mm/day C deg	Australian Scenario A1b Temp 2000 C deg	Australian Scenario A1b Temp 2099 C deg	Australian Scenario A1b Temp 2020-99 Change C deg
Malabar Transmitter Annex	1.9	4.4	2.5	22.1	16.4	-5.7	1.9	3.1	1.2	22.3	18.3	-4.0	1.0	2.6	1.6	20.1	19.6	-0.6
Mead Army National Guard Facility	0.8	1.3	0.6	-5.7	-4.3	1.4	0.8	0.7	0.0	-5.8	-1.5	4.3	0.6	0.9	0.3	-5.5	-6.5	-1.0
Milan Arsenal and Wildlife Management Area	9.6	3.2	-6.4	4.7	3.2	-1.5	9.6	1.1	-8.5	4.7	7.0	2.3	5.7	3.2	-2.4	3.0	5.1	2.1
Military Ocean Terminal Sunny Point	3.9	4.5	0.6	15.1	6.7	-8.4	3.9	2.7	-1.2	15.2	9.1	-6.1	4.6	5.1	0.5	8.2	9.6	1.5
Mount Baker Helicopter Training Area	17.6	11.4	-6.3	-3.0	-1.0	2.0	17.6	16.6	-1.0	-2.6	3.3	5.9	13.3	9.5	-3.8	-2.4	-0.2	2.2
Nap of the Earth Army Helicopter Training Area	16.1	12.9	-3.2	1.4	3.2	1.8	16.1	13.5	-2.6	1.7	5.8	4.1	7.7	8.0	0.3	1.3	3.5	2.2
Nap of the Earth Army Helicopter Training Area	8.3	6.3	-2.0	4.3	6.2	1.9	8.3	7.1	-1.2	4.6	9.1	4.5	4.5	4.2	-0.3	4.3	6.4	2.1
Natick Laboratories Military Reservation	5.9	4.3	-1.5	-0.4	1.0	1.4	5.9	4.0	-1.8	-0.6	-0.2	0.4	1.8	5.0	3.2	-6.0	-0.5	5.4
New Cumberland General Depot (US Military Reservation)	5.8	3.0	-2.8	2.1	1.5	-0.7	5.8	2.9	-2.9	2.2	1.5	-0.7	1.8	3.4	1.6	-2.1	-0.1	2.1
Newport Army Ammunition Plant	6.8	3.1	-3.7	-0.7	-1.0	-0.3	6.8	3.0	-3.8	-0.8	3.7	4.4	1.9	4.6	2.6	-2.6	-2.3	0.3
Picatinny Arsenal	7.7	3.9	-3.8	-0.3	-0.1	0.3	7.7	3.6	-4.1	-0.2	-0.8	-0.6	2.1	5.3	3.2	-4.2	-1.6	2.5
Pine Bluff Arsenal	7.3	2.2	-5.0	6.7	7.0	0.3	7.3	0.8	-6.5	6.8	10.1	3.3	5.0	2.4	-2.7	6.2	8.4	2.2
Presidio of Monterey	1.7	1.5	-0.1	10.2	12.4	2.2	1.7	7.4	5.7	10.2	14.5	4.2	1.5	9.5	7.9	9.9	13.6	3.7
Radford Army Ammunition Plant	4.0	2.2	-1.8	8.8	0.3	-8.5	4.0	2.0	-2.0	8.9	4.2	-4.6	2.5	2.0	-0.4	0.9	2.8	2.0
Radford Army Ammunition Plant	4.1	2.3	-1.8	8.6	0.4	-8.3	4.1	2.0	-2.1	8.7	4.2	-4.5	2.5	2.1	-0.4	0.8	2.7	1.9
Ravenna Arsenal	5.0	3.0	-2.0	0.5	-1.3	-1.9	5.0	4.0	-0.9	0.4	0.5	0.1	2.4	2.6	0.2	-4.2	-2.4	1.8
Red River Army Depot	3.9	1.4	-2.5	7.0	8.2	1.1	3.9	1.1	-2.8	7.2	11.8	4.6	4.1	1.5	-2.6	6.1	9.3	3.3
Redstone Arsenal	8.7	3.1	-5.6	10.1	4.4	-5.7	8.7	1.7	-7.0	10.1	8.5	-1.6	7.8	2.2	-5.6	5.6	7.6	2.0
Rock Island Arsenal	2.2	2.5	0.3	-5.4	-3.0	2.4	2.2	1.9	-0.3	-5.3	0.9	6.2	1.1	2.4	1.3	-5.2	-6.5	-1.3
Savanna Army Depot (Scheduled to close)	1.9	2.5	0.6	-6.3	-4.4	1.9	1.9	2.0	0.0	-6.3	-0.6	5.7	1.2	1.9	0.7	-7.0	-6.9	0.1
Seneca Army Depot (Scheduled to close)	3.2	2.4	-0.8	-2.8	-0.3	2.5	3.2	2.9	-0.3	-2.8	-0.8	2.1	1.2	1.7	0.6	-5.3	-2.5	2.8
Sharpe General Depot (Field Annex)	1.3	1.0	-0.3	7.1	9.3	2.3	1.3	4.2	2.9	7.0	11.5	4.5	0.9	5.6	4.7	7.0	10.8	3.9

Installation Name	Australian Scenario A1b Temp 20-99 Change C_deg	Australian Scenario A1b Temp 2099 C_deg	Australian Scenario A1b Temp 2000 C_deg	Australian Scenario A1b Prcp 20-99 Change mm/day C_deg	Australian Scenario A1b Prcp 2099 mm/day C_deg	Australian Scenario A1b Prcp 2000 mm/day	Canadian Scenario B1 Temp 20-99 Change C_deg	Canadian Scenario B1 Temp 2099 C_deg	Canadian Scenario B1 Temp 2000 C_deg	Canadian Scenario B1 Prcp 20-99 Change mm/day	Canadian Scenario B1 Prcp 2099 mm/day	Canadian Scenario B1 Prcp 2000 mm/day	Canadian Scenario A1b Temp 20-99 Change C_deg	Canadian Scenario A1b Temp 2099 C_deg	Canadian Scenario A1b Temp 2000 C_deg	Canadian Scenario A1b Prcp 20-99 Change mm/day	Canadian Scenario A1b Prcp 2099 mm/day	Canadian Scenario A1b Prcp 2000 mm/day
Sierra Army Depot	4.1	4.7	0.6	3.1	3.5	0.4	5.7	5.0	-0.7	1.1	2.5	1.4	3.5	2.7	-0.8	-0.6	0.7	1.4
Sierra Army Depot	4.0	2.6	-1.4	3.6	4.1	0.5	5.8	2.9	-2.8	1.3	2.9	1.6	3.6	0.6	-3.0	-0.7	0.9	1.6
Snoqualmie National Forest	3.0	0.5	-2.5	0.8	7.2	6.4	5.7	3.8	-1.9	-4.7	10.5	15.1	2.5	0.5	-1.9	-4.8	10.4	15.1
Sunflower Army Ammunition Plant	-2.1	-2.0	0.1	0.6	1.4	0.8	5.8	4.9	-0.8	-0.9	0.4	1.3	0.9	0.3	-0.6	-0.2	1.1	1.3
Tooele Army Depot	4.0	0.1	-3.9	3.2	3.7	0.6	7.3	2.3	-5.0	0.4	1.6	1.2	3.9	-1.5	-5.4	-0.4	0.9	1.2
Tooele Army Depot	3.9	2.5	-1.4	2.2	2.6	0.4	7.3	4.8	-2.5	0.2	1.1	0.9	3.7	0.8	-2.9	-0.3	0.6	0.9
US Army Aberdeen Proving Ground	1.8	2.1	0.3	2.4	4.5	2.1	-2.8	3.3	6.1	-3.5	2.5	6.0	-3.5	2.5	6.0	-3.1	3.0	6.0
US Army Ammunition Depot	3.5	7.0	3.5	-1.0	0.9	2.0	7.1	9.8	2.8	-1.6	0.5	2.1	3.5	5.9	2.4	-1.3	0.8	2.1
US Army Reserve Center	5.3	0.8	-4.5	3.4	5.5	2.1	0.0	0.9	1.0	-1.9	4.2	6.1	1.1	2.2	1.1	-1.3	4.8	6.1
US Garrison, Fort Detrick	1.6	1.3	-0.3	1.5	3.5	2.0	-2.3	2.9	5.2	-3.1	2.2	5.3	-3.3	1.8	5.1	-2.8	2.5	5.3
Utah Launch Complex White Sands Missile	3.8	-1.8	-5.6	1.1	1.3	0.2	6.5	-0.5	-7.1	0.3	0.6	0.3	5.1	-2.6	-7.7	0.0	0.3	0.3
Warrenton Training Center Military Reservation	1.5	2.4	0.9	1.1	3.3	2.2	-2.9	4.0	6.8	-3.0	2.0	5.0	-4.4	2.4	6.8	-2.5	2.6	5.0
Warrenton Training Center Military Reservation	1.4	1.8	0.3	1.2	3.4	2.2	-2.7	3.4	6.2	-3.0	2.1	5.1	-4.2	1.9	6.1	-2.5	2.6	5.1
West Point US Military Academy	3.0	-1.6	-4.6	3.2	5.2	2.0	-0.1	-0.7	-0.6	-3.7	3.7	7.4	0.7	0.1	-0.5	-3.4	4.0	7.4
White Sands Missile Range	6.4	6.0	-0.4	0.1	0.6	0.5	3.8	4.2	0.4	0.4	0.5	0.0	3.2	3.6	0.4	0.1	0.2	0.0
White Sands Missile Range	6.3	7.6	1.3	0.0	0.5	0.5	3.8	6.0	2.3	0.4	0.4	0.1	3.1	5.4	2.2	0.1	0.2	0.1
White Sands Missile Range	6.3	8.5	2.2	0.2	0.7	0.4	3.7	6.8	3.1	0.5	0.5	0.1	3.1	6.1	3.0	0.1	0.2	0.1
Yakima Firing Center	3.6	-0.9	-4.5	0.2	1.0	0.8	6.3	2.5	-3.8	-0.9	1.3	2.2	2.8	-1.0	-3.7	-0.9	1.3	2.2
Yuma Proving Ground	4.2	14.9	10.7	0.5	1.0	0.5	4.9	13.9	9.0	0.8	0.9	0.1	4.4	14.1	9.6	0.1	0.2	0.1

Table A2. Ecosystems associates with Army installations.

Installation Name	Bailey Ecosystem Name	Omernik Ecosystem Name	Omernik Ecosystem Name
Aberdeen Proving Ground Military Reservation	Interior Low Plateau, Highland Rim like	Interior Plateau like	Ozark Highlands like
Anniston Army Depot	Southern Appalachian Piedmont like	Southeastern Plains like	Ozark Highlands like
Arlington National Cemetery	Mid Coastal Plains, Western like	South Central Plains like	Ozark Highlands like
Army Chemical Center	Interior Low Plateau, Highland Rim like	Interior Plateau like	Ozark Highlands like
Army Reserve Outdoor Training Area	Central Dissected Till Plains like	Central Basin and Range like	Northern Basin and Range like
Army Training Area	South-Central Great Plains like	Central Great Plains like	Northern Basin and Range like
Badger Army Ammunition Plant	Central Dissected Till Plains like	Driftless Area like	Northern Basin and Range like
Bearmouth National Guard Training Area	North Central US Driftless and Escarpment like	Driftless Area like	Western Corn Belt Plains like
Belle Mead General Depot	North Central Glaciated Plains like	Eastern Corn Belt Plains like	Ozark Highlands like
Blossom Point Field Test Facility	Central Till Plains, Beech-Maple like	South Central Plains like	Ozark Highlands like
Buckeye National Guard Target Range	Mid Coastal Plains, Western like	Chihuahuan Deserts like	Sonoran Basin and Range like
Buckley Air National Guard AF Base	Basin and Range like	Central Great Plains like	Central Great Plains like
Camden Test Annex	South-Central Great Plains like	Northeastern Highlands like	Southern Michigan/Northern Indiana Drift Plains like
Camp Adair Military Reservation	Northern Glaciated Allegheny Plateau like	Eastern Corn Belt Plains like	Interior Plateau like
Camp Atterbury Military Reservation	Central Till Plains, Beech-Maple like	Eastern Corn Belt Plains like	Interior River Valleys and Hills like
Camp Bullis	Central Till Plains, Beech-Maple like	Edwards Plateau like	Sonoran Basin and Range like
Camp Dodge Military Reservation	Basin and Range like	Western Corn Belt Plains like	Northern Basin and Range like
Camp Grayling Military Reservation	North-Central Glaciated Plains like	Northern Lakes and Forests like	Central Corn Belt Plains like
Camp Johnson	Northern Great Lakes like	Central Corn Belt Plains like	Central Irregular Plains like
Camp Joseph T. Robinson	Central Loess Plains like	Interior Plateau like	East Central Texas Plains like
Camp MacKall Military Reservation	Coastal Plains and Flatwoods, Lower like	Southeastern Plains like	East Central Texas Plains like
Camp Parks Military Reservation	Oak Woods and Prairies like	East Central Texas Plains like	Southern Coastal Plain
Camp Riley Military Reservation	Interior Low Plateau, Highland Rim like	Interior Plateau like	Interior Plateau like
Camp Roberts Military Reservation	Basin and Range like	Edwards Plateau like	Edwards Plateau like
Camp Swift N. G. Facility	Rio Grande Plain like	Southern Coastal Plain like	Southern Texas Plains like
Camp Williams	North Central US Driftless and Escarpment like	Driftless Area like	Northern Lakes and Forests like
Camp Williams	North-Central Glaciated Plains like	Western Corn Belt Plains like	Central Great Plains like

Installation Name	Bailey Ecosystem Name	Omernik Ecosystem Name	Bailey Ecosystem Name	Omernik Ecosystem Name
Charles Melvin Price Support Center	Central Loess Plains like	Interior River Valleys and Hills like	Ozark Highlands like	Ozark Highlands like
Cornhusker Army Ammunition Plant	South-Central Great Plains like	Central Great Plains like	Central Dissected Till Plains like	Northern Basin and Range like
Custer Reserve Forces Training Area	South Central Great Lakes like	Southern Michigan/Northern Indiana Drift Plains like	Central Loess Plains like	Interior River Valleys and Hills like
Dugway Proving Grounds	South-Central Great Plains like	Central Great Plains like	Texas High Plains like	High Plains like
Edgewood Arsenal	Interior Low Plateau, Highland Rim like	Interior Plateau like	Ozark Highlands like	Ozark Highlands like
Florence Military Reservation	Basin and Range like	Chihuahuan Deserts like	Rio Grande Plain like	Southern Texas Plains like
Fort A. P. Hill Military Reservation	Mid Coastal Plains, Western like	South Central Plains like	Ozark Highlands like	Ozark Highlands like
Fort Belvoir Military Reservation	Mid Coastal Plains, Western like	South Central Plains like	Ozark Highlands like	Ozark Highlands like
Fort Benning Military Reservation	Southern Appalachian Piedmont like	Southeastern Plains like	Oak Woods and Prairies like	East Central Texas Plains like
Fort Bliss	Basin and Range like	Chihuahuan Deserts like	Basin and Range like	Chihuahuan Deserts like
Fort Bliss McGregor Range	Basin and Range like	Chihuahuan Deserts like	Basin and Range like	Chihuahuan Deserts like
Fort Bragg Military Reservation	Coastal Plains and Flatwoods, Lower like	Southeastern Plains like	Oak Woods and Prairies like	East Central Texas Plains like
Fort Campbell	Interior Low Plateau, Highland Rim like	Interior Plateau like	Ozark Highlands like	Ozark Highlands like
Fort Carson Military Reservation	Arkansas Tablelands like	Arizona/New Mexico Plateau like	South-Central Great Plains like	Central Great Plains like
Fort Carson Military Reservation	Arkansas Tablelands like	Arizona/New Mexico Plateau like	Texas High Plains like	High Plains like
Fort Dix Military Reservation	Interior Low Plateau, Highland Rim like	Interior Plateau like	Ozark Highlands like	Ozark Highlands like
Fort Drum	Northern Glaciated Allegheny Plateau like	Northeastern Highlands like	Central Loess Plains like	Interior River Valleys and Hills like
Fort Ethan Allen Military Reservation	Central Loess Plains like	Central Corn Belt Plains like	Central Loess Plains like	Interior River Valleys and Hills like
Fort Eustis Military Reservation	Southern Appalachian Piedmont like	Southeastern Plains like	Ozark Highlands like	Ozark Highlands like
Fort George G. Meade	Mid Coastal Plains, Western like	South Central Plains like	Ozark Highlands like	Ozark Highlands like
Fort Gillem Heliport	Southern Appalachian Piedmont like	Southeastern Plains like	Ozark Highlands like	Ozark Highlands like
Fort Gordon	Coastal Plains and Flatwoods, Lower like	Southeastern Plains like	Mid Coastal Plains, Western like	South Central Plains like
Fort Hood	Cross Timbers and Prairie like	Cross Timbers like	Rio Grande Plain like	Southern Texas Plains like
Fort Huachuca	Basin and Range like	Chihuahuan Deserts like	Rio Grande Plain like	Southern Texas Plains like
Fort Irwin	Texas High Plains like	High Plains like	Stockton Plateau like	Sonoran Basin and Range like
Fort Jackson	Coastal Plains and Flatwoods, Lower like	Southeastern Plains like	Mid Coastal Plains, Western like	South Central Plains like
Fort Knox	Interior Low Plateau, Highland Rim like	Interior Plateau like	Ozark Highlands like	Ozark Highlands like
Fort Leavenworth Military Reservation	Columbia Basin like	Central Irregular Plains like	Columbia Basin like	Central Irregular Plains like
Fort Lee Military Reservation	Southern Appalachian Piedmont like	Southeastern Plains like	Ozark Highlands like	Ozark Highlands like

Installation Name	Bailey Ecosystem Name	Omernik Ecosystem Name	Omernik Ecosystem Name
Fort Leonard Wood Military Reservation	Ozark Highlands like	Columbia Basin like	Central Irregular Plains like
Fort Lewis Military Reservation	Interior Low Plateau, Highland Rim like	Interior Plateau like	Interior Plateau like
Fort MacArthur	Basin and Range like	Rio Grande Plain like	Southern Texas Plains like
Fort McCoy	North Central US Driftless and Escarpment like	North Central US Driftless and Escarpment like	Driftless Area like
Fort McPherson	Southern Appalachian Piedmont like	Ozark Highlands like	Ozark Highlands like
Fort Monmouth Military Reservation	Interior Low Plateau, Highland Rim like	Interior Plateau like	Ozark Highlands like
Fort Monroe Military Reservation	Southern Appalachian Piedmont like	Southeastern Plains like	Ozark Highlands like
Fort Polk Military Reservation	Mid Coastal Plains, Western like	Coastal Plains and Flatwoods, Lower like	Southern Coastal Plain
Fort Riley Military Reservation	South-Central Great Plains like	South Central Plains like	Central Great Plains like
Fort Ritchie Raven Rock Site	Ozark Highlands like	Ridge and Valley like	Ozark Highlands like
Fort Rucker Military Reservation	Coastal Plains and Flatwoods, Lower like	Oak Woods and Prairies like	East Central Texas Plains like
Fort Sill Military Reservation	Cross Timbers and Prairie like	Texas High Plains like	High Plains like
Fort Stewart	Coastal Plains and Flatwoods, Lower like	Mid Coastal Plains, Western like	South Central Plains like
Fort William H. Harrison Military Reservation	North-Central Glaciated Plains like	North-Central Glaciated Plains like	Western Corn Belt Plains like
Fort Wolters	Cross Timbers and Prairie like	Stockton Plateau like	Sonoran Basin and Range like
Globecom Radio Receiving Station	Mid Coastal Plains, Western like	Ozark Highlands like	Ozark Highlands like
Greencastle Military Reservation	Ozark Highlands like	Ridge and Valley like	Ozark Highlands like
Hunter Army Airfield	Coastal Plains and Flatwoods, Lower like	Mid Coastal Plains, Western like	South Central Plains like
Hunter-Liggett Military Reservation	Cross Timbers and Prairie like	Coastal Plains and Flatwoods, Lower like	Southern Coastal Plain
Hunter-Liggett Military Reservation	Cross Timbers and Prairie like	Coastal Plains and Flatwoods, Lower like	Southern Coastal Plain
Iowa Army Ammunition Plant	Central Loess Plains like	Central Dissected Till Plains like	Central Irregular Plains like
Joliet Army Ammunition Plant	Central Loess Plains like	Central Loess Plains like	Interior River Valleys and Hills like
Joliet Army Ammunition Plant	Central Loess Plains like	Central Dissected Till Plains like	Central Irregular Plains like
Kansas Army Ammunition Plant	Columbia Basin like	Cross Timbers and Prairie like	Central Great Plains like
Kearney Rifle Range	South-Central Great Plains like	Central Dissected Till Plains like	Northern Basin and Range like
Lake City Army Ammunition Plant	Columbia Basin like	Columbia Basin like	Central Irregular Plains like
LaPorte Outdoor Training Facility	South Central Great Lakes like	Central Loess Plains like	Southern Michigan/Northern Indiana Drift Plains like	Interior River Valleys and Hills like
Letterkenny Army Depot	Ozark Highlands like	Ozark Highlands like	Ridge and Valley like	Ozark Highlands like
Longhorn Ordnance Army Ammo Plant	Mid Coastal Plains, Western like	Coastal Plains and Flatwoods, Lower like	South Central Plains like	Southern Coastal Plain

Installation Name	Bailey Ecosystem Name	Omernik Ecosystem Name	Omernik Ecosystem Name
Los Alamitos Armed Forces Reserve Center	Basin and Range like	Southern Texas Plains like	Southern Texas Plains like
Louisiana Ordnance Plant	Mid Coastal Plains, Western like	South Central Plains like	Southern Coastal Plain
Malabar Transmitter Annex	Coastal Plains and Flatwoods, Lower like	Southern Coastal Plain like	Southeastern Plains like
Mead Army National Guard Facility	Central Dissected Till Plains like	Central Basin and Range like	Northern Basin and Range like
Milan Arsenal and Wildlife Management Area	Interior Low Plateau, Highland Rim like	Interior Plateau like	Ozark Highlands like
Military Ocean Terminal Sunny Point	Coastal Plains and Flatwoods, Lower like	Southeastern Plains like	South Central Plains like
Mount Baker Helicopter Training Area	Central Till Plains, Beech-Maple like	Eastern Corn Belt Plains like	Eastern Corn Belt Plains like
Nap of the Earth Army Helicopter Training Area	Interior Low Plateau, Highland Rim like	Interior Plateau like	Interior Plateau like
Nap of the Earth Army Helicopter Training Area	Central Till Plains, Beech-Maple like	Eastern Corn Belt Plains like	Eastern Corn Belt Plains like
Natick Laboratories Military Reservation	Ozark Highlands like	Ridge and Valley like	Ridge and Valley like
New Cumberland General Depot (US Military Reservation)	Ozark Highlands like	Ridge and Valley like	Ozark Highlands like
Newport Army Ammunition Plant	Central Till Plains, Beech-Maple like	Eastern Corn Belt Plains like	Interior River Valleys and Hills like
Picatinny Arsenal	Central Till Plains, Beech-Maple like	Eastern Corn Belt Plains like	Ridge and Valley like
Pine Bluff Arsenal	Interior Low Plateau, Highland Rim like	Interior Plateau like	East Central Texas Plains like
Presidio of Monterey	Coastal Plains and Flatwoods, Lower like	Southern Coastal Plain like	Southern Coastal Plain
Radford Army Ammunition Plant	Mid Coastal Plains, Western like	South Central Plains like	Ozark Highlands like
Radford Army Ammunition Plant	Mid Coastal Plains, Western like	South Central Plains like	Ozark Highlands like
Ravenna Arsenal	Ozark Highlands like	Ridge and Valley like	Interior River Valleys and Hills like
Red River Army Depot	Mid Coastal Plains, Western like	South Central Plains like	Southern Coastal Plain
Redstone Arsenal	Southern Appalachian Piedmont like	Southeastern Plains like	Ozark Highlands like
Rock Island Arsenal	Central Dissected Till Plains like	Northern Basin and Range like	Central Irregular Plains like
Savanna Army Depot (Scheduled to close)	Central Dissected Till Plains like	Northern Basin and Range like	Central Corn Belt Plains like
Seneca Army Depot (Scheduled to close)	Central Loess Plains like	Interior River Valleys and Hills like	Ozark Highlands like
Sharpe General Depot (Field Annex)	Cross Timbers and Prairie like	Cross Timbers like	Edwards Plateau like

Installation Name	Bailey Ecosystem Name	Omernik Ecosystem Name	Omernik Ecosystem Name
Sierra Army Depot	Columbia Basin like	Central Irregular Plains like	Central Great Plains like
Sierra Army Depot	Central Dissected Till Plains like	Northern Basin and Range like	Central Great Plains like
Snoqualmie National Forest	Central Till Plains, Beech-Maple like	Eastern Corn Belt Plains like	Eastern Corn Belt Plains like
Sunflower Army Ammunition Plant	Columbia Basin like	Central Irregular Plains like	Central Irregular Plains like
Tooele Army Depot	North-Central Glaciated Plains like	Western Corn Belt Plains like	Central Great Plains like
Tooele Army Depot	South-Central Great Plains like	Central Great Plains like	Central Great Plains like
US Army Aberdeen Proving Ground	Interior Low Plateau, Highland Rim like	Interior Plateau like	Ozark Highlands like
US Army Ammunition Depot	Edwards Plateau like	Ozark Highlands like	Edwards Plateau like
US Army Reserve Center	Ozark Highlands like	Ridge and Valley like	Ridge and Valley like
US Garrison, Fort Detrick	Mid Coastal Plains, Western like	South Central Plains like	Ozark Highlands like
Utah Launch Complex White Sands Missile	Northwestern Great Plains like	High Plains like	Arizona/New Mexico Plateau like
Warrenton Training Center Military Reservation	Mid Coastal Plains, Western like	South Central Plains like	Ozark Highlands like
Warrenton Training Center Military Reservation	Mid Coastal Plains, Western like	South Central Plains like	Ozark Highlands like
West Point US Military Academy	Central Till Plains, Beech-Maple like	Eastern Corn Belt Plains like	Ridge and Valley like
White Sands Missile Range	Basin and Range like	Chihuahuan Deserts like	Chihuahuan Deserts like
White Sands Missile Range	Basin and Range like	Chihuahuan Deserts like	Chihuahuan Deserts like
White Sands Missile Range	Pecos Valley like	Arizona/New Mexico Plateau like	Chihuahuan Deserts like
Yakima Firing Center	Central Dissected Till Plains like	Northern Basin and Range like	Central Irregular Plains like
Yuma Proving Ground	Sonoran Mojave Desert like	Sonoran Basin and Range like	Sonoran Basin and Range like

Appendix B: Data Used To Generate Ecosystem Change Evaluations in Sections 6.4.3

Percent Habit Unchanged By Climate Change

Installation	Size (0.02 X 0.02)-degree cells	PCM Model				Hadley Model			
		Low Emissions		High Emissions		Low Emissions		High Emissions	
		PCM B1 2050	PCM B1 2100	PCM A1 2050	PCM A1 2100	HAD B1 2050	HAD B1 2100	HAD A1 2050	HAD A1 2100
Camp Adair Military Reservation	2	100	100	100	100	100	100	100	100
Hunter Army Airfield	77	100	100	100	100	100	100	100	21
Fort Stewart	2294	100	97	100	95	100	99	100	14
Arlington National Cemetery	4	100	100	100	100	100	100	100	0
Army Reserve Outdoor Training Area	4	100	100	100	100	100	100	100	0
Army Training Area	20	100	100	100	100	100	100	100	0
Globecom Radio Receiving Station	15	100	100	100	100	100	100	100	0
Kearney Rifle Range	8	100	100	100	100	100	100	100	0
LaPorte Outdoor Training Facility	6	100	100	100	100	100	100	100	0
Malabar Transmitter Annex	4	100	100	100	100	100	100	100	0
US Army Reserve Center	4	100	100	100	100	100	100	100	0
Florence Military Reservation	56	100	100	89	79	89	89	89	63
Camp Grayling Military Reservation	2451	97	97	97	97	97	94	97	13
Savanna Army Depot (Scheduled to close)	228	100	100	100	95	97	95	95	0
Fort Irwin	5112	91	91	95	88	90	86	92	46
Fort Belvoir Military Reservation	110	100	100	97	90	90	69	90	29
Camp Roberts Military Reservation	425	95	91	100	97	91	84	89	10
Fort Detrick	8	100	100	100	100	100	50	100	0
Badger Army Ammunition Plant	35	100	100	100	100	83	83	83	0
Fort William H. Harrison Military Reservation	30	100	100	100	100	100	70	70	0
Fort George G. Meade	56	100	100	100	79	100	79	79	0
Cornhusker Army Ammunition Plant	8	100	100	100	100	100	100	25	0
Bearmouth National Guard Training Area	15	100	100	100	73	100	67	67	13
Joliet Army Ammunition Plant	182	100	100	100	52	100	77	88	0
Custer Reserve Forces Training Area	96	100	100	100	51	100	78	78	0
Fort Riley Military Reservation	1280	98	98	98	98	95	54	60	0
Buckeye National Guard Target Range	10	100	100	100	0	100	100	100	0
Fitzsimons Army Medical Center (Closed)	4	100	50	50	50	100	100	100	50
Natick Laboratories Military Reservation	36	100	100	100	0	100	100	100	0
Louisiana Ordnance Plant	95	92	92	92	75	75	62	75	0
Nap of the Earth Army Helicopter Training Are	5338	75	73	73	71	72	71	70	50

Legend: 80-100% unchanged | 50-80% unchanged | 0-50% unchanged

| Percent Habit Unchanged By Climate Change | | PCM Model | | | | Hadley Model | | | |
| | | Low Emissions | | High Emissions | | Low Emissions | | High Emissions | |
Installation	Size (0.02 X 0.02)-degree cells	PCM B1 2050	PCM B1 2100	PCM A1 2050	PCM A1 2100	HAD B1 2050	HAD B1 2100	HAD A1 2050	HAD A1 2100
Camp Dodge Military Reservation	25	100	100	100	32	24	92	92	0
Fort Pickett Military Reservation (Closed)	352	97	97	88	91	91	18	53	0
Sharpe General Depot (Field Annex)	3	100	100	100	100	67	67	0	0
Fort McCoy	713	100	100	100	98	50	47	34	0
Newport Army Ammunition Plant	24	100	100	100	100	75	25	25	0
Dugway Proving Grounds	3640	86	78	77	63	71	67	69	8
US Army Aberdeen Proving Ground	205	82	77	73	67	62	67	62	27
Aberdeen Proving Ground Military Reservation	445	89	78	69	66	57	66	64	22
Camp Bullis	234	100	100	100	48	60	55	48	0
Edgewood Arsenal	40	70	70	70	70	70	70	70	20
New Cumberland General Depot (US Military R	130	88	88	78	54	66	54	66	8
Milan Arsenal And Wildlife Management Area	234	97	84	84	91	74	12	59	0
Fort Wolters	25	100	100	100	36	52	52	52	0
Fort Lewis Military Reservation	3089	61	61	64	54	60	56	60	40
Charles Melvin Price Support Center	8	100	100	100	100	25	0	25	0
Fort Devens (Closed)	140	66	66	66	50	66	66	66	0
Radford Army Ammunition Plant	240	90	87	97	70	27	27	27	19
Camp Swift N. G. Facility	650	100	100	100	37	31	24	29	8
Yuma Proving Ground	6052	84	74	81	15	73	42	52	1
Fort Rucker Military Reservation	702	74	74	74	74	48	43	28	0
Fort Benjamin Harrison (Closed)	35	100	100	100	100	0	0	0	0
Los Alamitos Armed Forces Reserve Center	8	100	100	100	100	0	0	0	0
Sacramento Army Depot (Closed)	4	50	50	100	50	50	50	50	0
Fort Carson Military Reservation	30777	62	57	57	26	57	55	71	11
Rock Island Arsenal	5	100	40	40	0	0	0	100	0
Kansas Army Ammunition Plant	44	100	100	100	73	0	0	0	0
Fort McClellan Military Reservation (Closed)	242	93	85	93	37	30	7	21	0
Fort A. P. Hill Military Reservation	728	65	48	89	72	33	23	33	0
Navajo Army Depot (Closed)	221	69	64	67	20	44	34	43	16
Camp Atterbury Military Reservation	198	58	58	60	57	57	34	32	0
Redstone Arsenal	273	96	44	100	38	27	23	23	0

Legend: 80-100% unchanged   50-80% unchanged   0-50% unchanged

Percent Habit Unchanged By Climate Change

| Installation | Size (0.02 X 0.02)-degree cells | PCM Model | | | | Hadley Model | | | |
| | | Low Emissions | | High Emissions | | Low Emissions | | High Emissions | |
		PCM B1 2050	PCM B1 2100	PCM A1 2050	PCM A1 2100	HAD B1 2050	HAD B1 2100	HAD A1 2050	HAD A1 2100
Fort Leonard Wood Military Reservation	756	100	100	100	12	15	0	7	0
White Sands Missile Range	13522	52	51	45	29	46	47	40	24
Camp Joseph T. Robinson	288	82	74	72	24	24	24	24	6
Red River Army Depot	242	59	51	51	51	54	23	31	0
Fort Polk Military Reservation	3450	94	93	90	18	10	4	7	0
Fort Wingate Depot Activity (Closed)	144	83	77	77	0	49	15	14	0
Fort Bragg Military Reservation	1392	75	52	32	45	31	50	30	1
Lake City Army Ammunition Plant	18	100	100	100	6	6	0	0	0
Camp Johnson	9	44	44	44	44	44	22	44	22
Fort Leavenworth Military Reservation	63	87	100	100	0	16	0	0	0
Fort Hood	4321	72	72	73	22	22	21	21	0
Fort Gillem Heliport	15	100	100	100	0	0	0	0	0
Fort McPherson	4	100	100	100	0	0	0	0	0
Fort Monmouth Military Reservation	8	100	100	100	0	0	0	0	0
Longhorn Ordnance Army Ammo Plant	42	100	100	100	0	0	0	0	0
Fort Benning Military Reservation	1599	78	48	81	38	30	11	14	0
Mount Baker Helicopter Training Area	8017	42	42	39	34	40	36	36	25
Buckley Air National Guard AF Base	30	30	30	30	0	100	50	50	0
US Army Ammunition Depot	169	98	89	100	2	0	0	0	0
Military Ocean Terminal Sunny Point	150	100	100	80	0	0	0	0	0
West Point US Military Academy	121	61	56	68	23	50	10	10	0
Fort Gordon	792	78	64	80	12	19	13	13	0
Fort Campbell	925	96	74	82	4	19	0	1	0
Fort Sill Military Reservation	900	42	38	38	38	38	38	38	0
Pine Bluff Arsenal	288	89	83	87	3	0	0	0	0
Fort Bliss McGregor Range	11190	35	31	29	35	43	44	23	12
Sunflower Army Ammunition Plant	35	83	83	83	0	0	0	0	0
Anniston Army Depot	90	70	53	70	20	11	11	11	0
Umatilla Chemical Depot (Closed)	88	59	55	27	23	24	20	24	0
Warrenton Training Center Military Reservation	130	47	25	64	18	31	18	22	0
Fort Lee Military Reservation	66	36	50	50	27	50	0	0	0

Legend: 80-100% unchanged | 50-80% unchanged | 0-50% unchanged

Percent Habitat Unchanged By Climate Change

Installation	Size (0.02 X 0.02)-degree cells	PCM Model				Hadley Model			
		Low Emissions		High Emissions		Low Emissions		High Emissions	
		PCM B1 2050	PCM B1 2100	PCM A1 2050	PCM A1 2100	HAD B1 2050	HAD B1 2100	HAD A1 2050	HAD A1 2100
Belle Mead General Depot	8	50	50	50	50	0	0	0	0
Fort Knox	2311	39	39	40	22	25	12	16	0
Iowa Army Ammunition Plant	170	72	55	36	7	7	7	7	0
Fort Jackson	527	64	13	13	54	13	13	13	0
Fort Ritchie Military Reservation (Closed)	2011	27	25	27	19	19	23	22	8
Hunter-Liggett Military Reservation	11581	23	22	23	22	21	19	21	11
Lexington-Blue Grass Army Depot (Closed)	1915	37	36	37	4	4	7	5	0
Fort Dix Military Reservation	1529	20	20	20	14	19	14	13	1
Fort Drum	3877	24	21	24	7	17	10	11	3
Fort Huachuca	6280	14	15	18	10	17	13	10	4
Army Chemical Center	4	0	0	100	0	0	0	0	0
Blossom Point Field Test Facility	15	100	0	0	0	0	0	0	0
Greencastle Military Reservation	9	0	0	0	0	0	100	0	0
Indiana Arsenal Army Ammunition Plant (Closed)	1439	15	15	15	10	10	9	11	0
Yakima Firing Center	12237	11	11	11	10	10	8	10	0
Ravenna Arsenal	1381	13	13	13	12	6	2	6	6
Tooele Army Depot	4069	9	9	7	6	6	4	4	3
Fort Bliss	8206	7	5	3	5	9	7	5	5
Seneca Army Depot (Scheduled to close)	1281	6	6	6	4	6	4	4	3
Fort Ord Military Reservation (Closed)	5071	6	5	5	6	6	5	7	1
Fort Eustis Military Reservation	1268	5	5	5	5	6	4	7	3
Fort Ethan Allen Military Reservation	1273	6	6	6	5	6	6	4	0
Fort Chaffee (Closed)	8316	7	7	7	4	5	1	3	0
Camp MacKall Military Reservation	66	12	9	0	0	0	12	0	0
Picatinny Arsenal	1305	6	5	6	3	5	0	0	0
Fort Indiantown Gap Military Reservation (Closed)	1513	8	5	5	1	1	1	2	1
Letterkenny Army Depot	1321	5	4	5	2	2	3	2	0
Pueblo Chemical Depot (Closed)	5344	2	2	2	2	2	2	3	0
Sierra Army Depot	7107	3	2	3	2	3	2	2	0
Camp Bonneville Military Reservation (Closed)	1226	2	2	2	2	2	2	3	0
Craney Island Disposal Area	1222	2	2	2	2	2	2	2	0

Legend: | 80-100% unchanged | 50-80% unchanged | 0-50% unchanged

| Percent Habit Unchanged By Climate Change | | PCM Model | | | | Hadley Model | | | |
| | | Low Emissions | | High Emissions | | Low Emissions | | High Emissions | |
Installation	Size (0.02 X 0.02)-degree cells	PCM B1 2050	PCM B1 2100	PCM A1 2050	PCM A1 2100	HAD B1 2050	HAD B1 2100	HAD A1 2050	HAD A1 2100
Fort Story Military Reservation	1220	1	1	1	1	1	1	1	0
Utah Launch Complex White Sands Missile	5301	2	1	3	1	1	1	1	0
Camp Parks Military Reservation	1221	2	0	0	0	0	0	0	0
Defense Depot Ogden (Closed)	1208	0	0	0	0	0	0	0	0
Fort Sheridan (Closed)	1202	0	0	0	1	0	0	0	0
Oakland Army Base (Closed)	3606	0	0	0	0	0	0	0	0
Presidio of Monterey	3609	0	0	0	0	0	0	0	0
Camden Test Annex	3	0	0	0	0	0	0	0	0
Fort Ritchie Raven Rock Site	9	0	0	0	0	0	0	0	0
Vint Hill Farms Station Military Reservation	4	0	0	0	0	0	0	0	0

Legend: | 80-100% unchanged | 50-80% unchanged | 0-50% unchanged

Appendix C: Maps Supporting Erosion Analysis

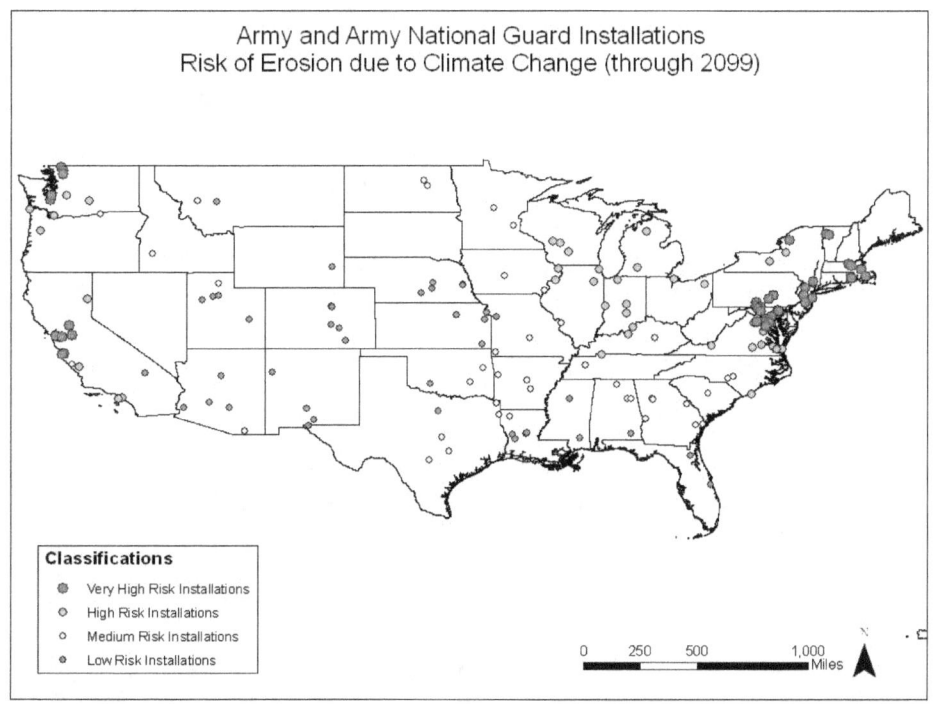

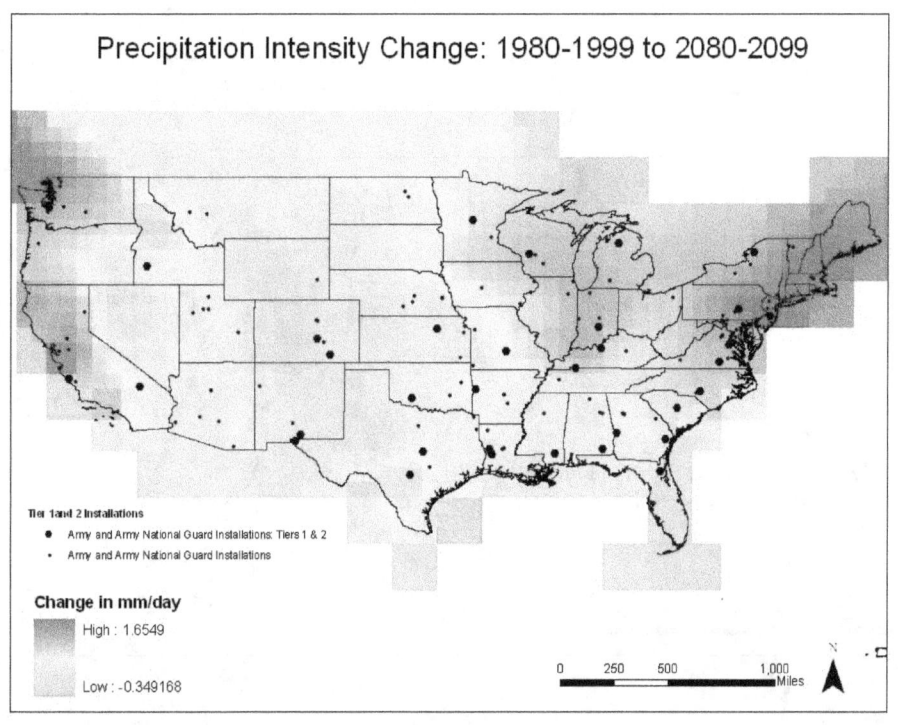

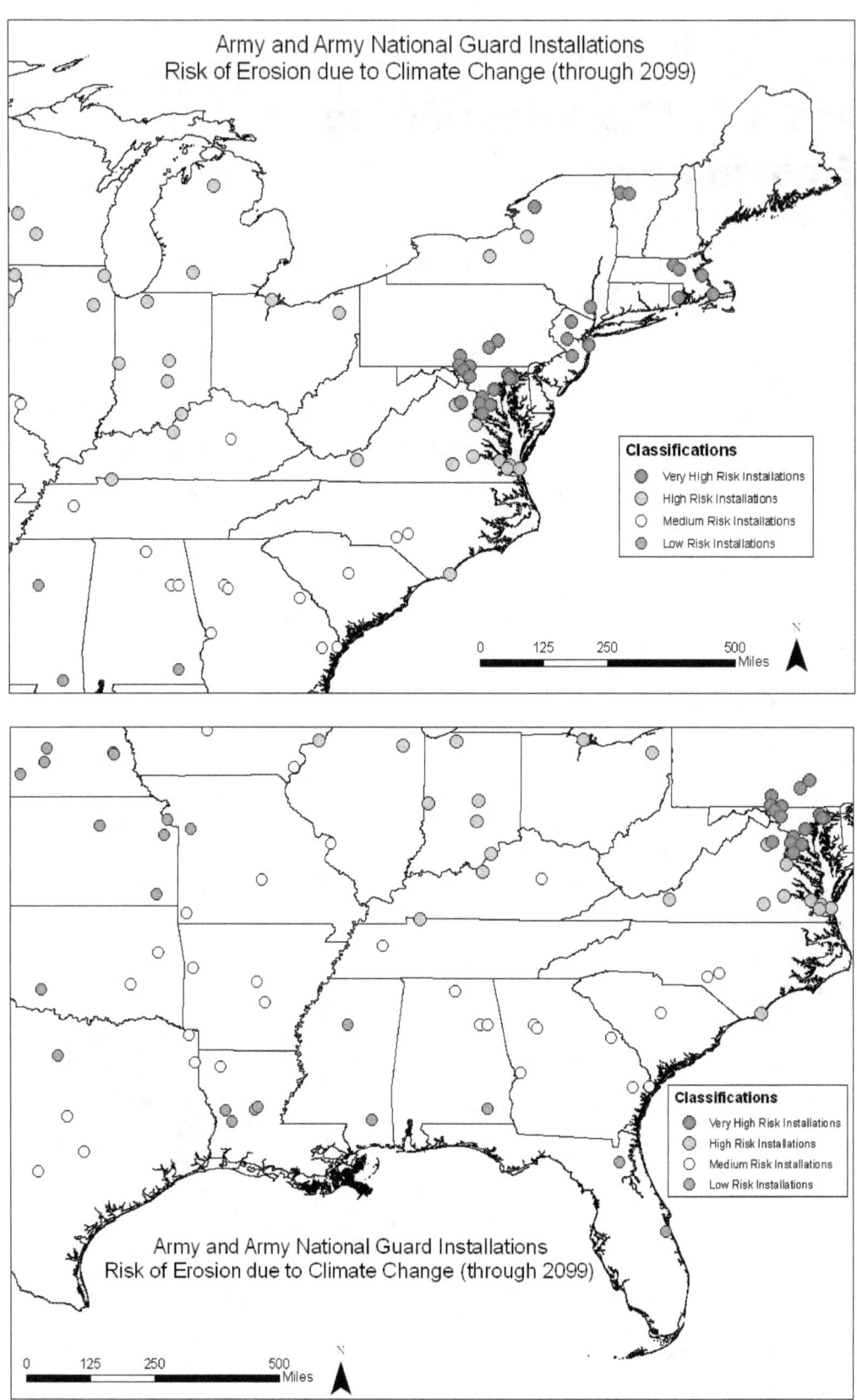

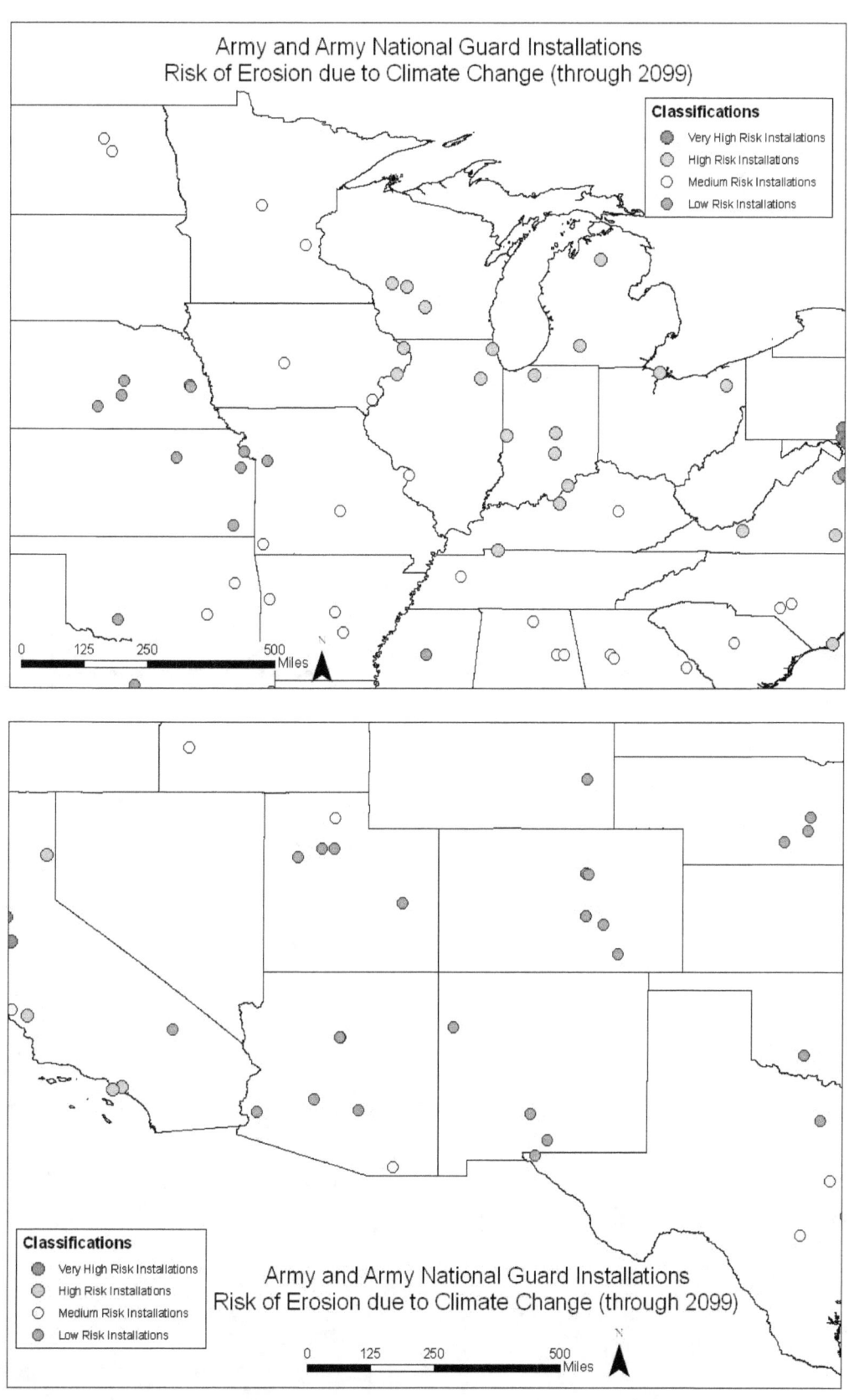

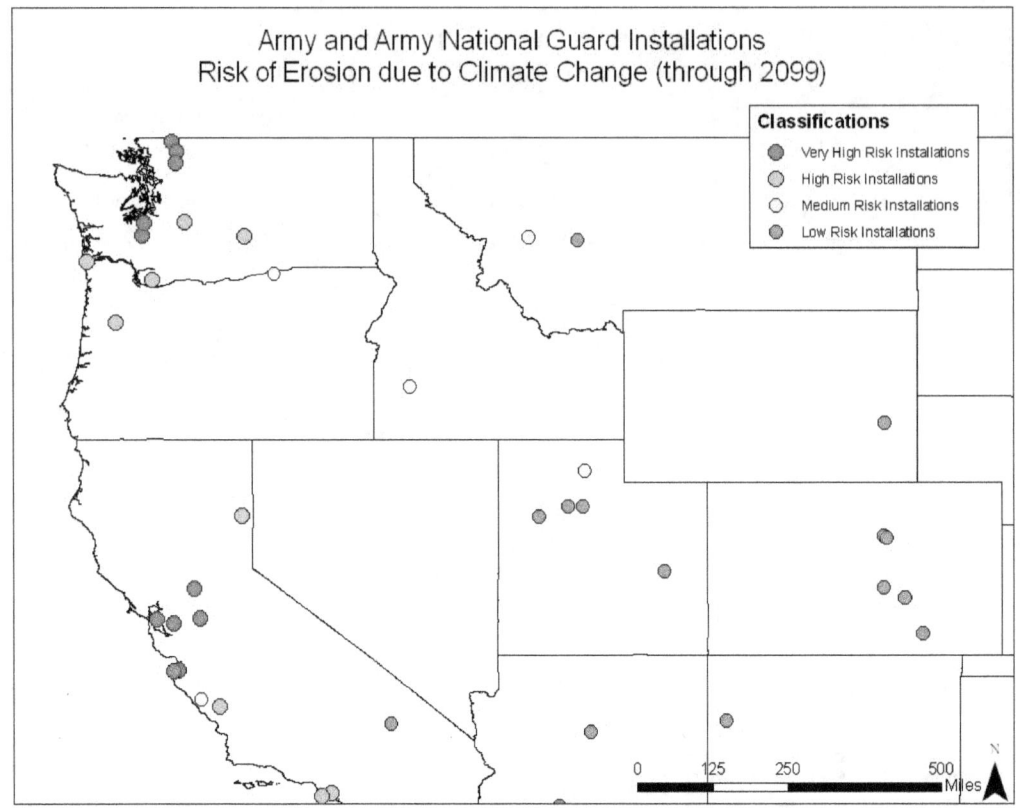

Appendix D: TES Habitat Statements Relating to Climate

For the Red-Cockaded Woodpecker (RCW):

- Young birds leave the nest after 26-29 days (Ligon 1970), and most nesting activity is finished by early July (Baker 1983a, Wood 1983b).
- Molting may become particularly heavy towards late summer and food requirements increase during these periods of heavy molt (Jackson 1983a). Post-juvenile plumage is attained by late fall or early winter of the first year (Jackson 1983a).
- RCW maintain territories throughout the year, and appear to recognize precise boundaries (Ligon 1970).
- Mean territory sizes of 70 ha were obtained (Hooper et al. 1982, Repasky 1984, Blue 1985).
- The relatively short dispersal distance implies that rates of inbreeding may be high even though close inbreeding is avoided (Walters 1990).
- Habitat consists of open, mature pine woodlands, and only rarely consists of deciduous or mixed pine-hardwoods located near pine woodlands (Steirly 1957, Hooper et al. 1980, US Fish and Wildlife Service 1980, Kalisz and Boettcher 1991).
- Optimal habitat is characterized as a broad savanna with a scattered overstory of large pines and a dense groundcover containing a diversity of grass, forb, and shrub species (Hooper et al. 1980, AOU 1991). Midstory vegetation is sparse or absent (Hooper et al. 1980, Locke et al. 1983, Hooper et al. 1991, Loeb et al. 1993).
- The open, park-like characteristic of the habitat is maintained by low intensity fires.
- Fire-maintained, old-growth pine savannas was once the dominant ecosystem in the Southeast.
- Roosting and nesting cavities have been found in longleaf, loblolly (*Pinus taeda*), shortleaf (*Pinus echinata*), slash (*Pinus ellioti*), pond pine (*Pinus rigida*), and even bald cypress (*Taxodium disthicus*) (Dennis 1971).
- In addition to requirements for old pine trees, appropriate habitat also includes open, park-like conditions extending across the area surrounding a cluster of cavity trees Walters (1991).

- Walters monitored a population of 12 cavity clusters for 9 years and found the population to be stable until logging cleared much of the foraging habitat in the area (pers. comm.).
- Areas that have suitable habitat characteristics, yet lack suitable cavity trees, will not likely be occupied by red-cockaded woodpeckers (Walters 1991).
- Birds drink water from flooded holes in trees and from the ground (Murphrey 1939, Hooper et al. 1980).
- The primary actions needed to accomplish delisting and downlisting recovery goals are: (1) application of frequent fire to both clusters and foraging habitat, (2) protection and development of large, mature pines throughout the landscape, (3) protection of existing cavities and judicious provisioning of artificial cavities, (4) provision of sufficient recruitment clusters in locations chosen to enhance the spatial arrangement of groups, and (5) restoration of sufficient habitat quality and quantity to support the large populations necessary for recovery (USFWS2003).
- Management centers on maintaining old-growth pine forests and establishing an effective prescribed burning program. Minimum tree ages should be 100–125 years for longleaf pine, 80–150 years for shortleaf, and 80–120 years for loblolly.
- Preserves large enough to support more than 25 active clusters will be stable for long periods of time and probably require infrequent intervention (so long as optimal habitat conditions are maintained). Preserves of approximately 1000–4000 ha (2470–9890 acres) have the capacity to support populations of this size.
- The slow rate of habitat colonization exhibited by red-cockaded woodpeckers (Walters 1990) implies that it will be difficult to re-establish extirpated populations.

For the Gopher Tortoise (GT):

- Urban development and agricultural conversion (including commercial forestry) are the primary threats.
- Any development that fragments a population and/or creates a barrier to the natural movement of gopher tortoises likely will negatively impact that population.
- Incompatible silvicultural practices affect the GT.
- A longer season of activity results in females maturing faster.
- Gopher tortoises excavate deep burrows that provide shelter from climate extremes.

- The high humidity associated with the burrow may offer the tortoise protection from desiccation.
- Gopher tortoises desiccate more rapidly when deprived of a burrow than any other member of the genus *Gopherus*.
- GT may withstand relatively high body temperatures.
- Critical thermal maximum is reported as 43.9 °C (111 °F).
- Individuals that may be forced to abandon isolated patches of habitat in areas surrounded by human dwellings seem doomed to perish.
- Individuals generally maintain a well-defined activity (home) range. A large GT may encompass up to 6+ ha over several years.
- The longest movement made was 0.74 km by an emigrating subadult.
- Commonly occupies habitats with a well-drained sandy substrate, ample herbaceous vegetation for food, and sunlit areas for nesting.
- GT prefers open habitats that support a wide variety of herbaceous ground cover vegetation for forage; it usually abandons densely canopied areas and frequently can be found in disturbed habitats such as roadsides, fence-rows, old fields, and the edges of overgrown (unburned) uplands.
- Upland habitats with extensive canopies reduce the amount of direct sunlight on the ground, which may hamper tortoises from reaching minimum thermal requirements for normal daily activities.
- Excessive shade decreases herbaceous vegetation essential for GT growth, development, and reproduction.
- GT temporarily abandons marginal habitats during periods of drought.
- Activities of gopher tortoises away from their burrows are limited in the winter months and increase as seasonal temperatures increase.
- GT prefers relatively open-canopied habitats that provide sunlit areas for nesting and thermoregulation, and ample herbaceous ground vegetation for forage.
- Landers and Speake (1980) recognized that gopher tortoises can be maintained on small management units, but they proposed that larger units (up to several hundred hectares) would lessen the impact of emigration and mortality.
- Gopher tortoises function as a "keystone species."
- Tortoise habitat quality may be viewed as a dynamic gradient.
- Highest GT densities occur in grassy, open-canopied sites (Auffenberg and Franz 1982, Mushinsky and McCoy 1994). Prescribed burning is the preferred method for managing gopher tortoise habitats.
- Sandhill habitat responds well to summer burns on a 2-7 year periodicity.

www.ingramcontent.com/pod-product-compliance
Lightning Source LLC
Chambersburg PA
CBHW080257180526
45167CB00006B/2561